Modelling of Environmental Chemical Exposure and Risk

NATO Science Series

A Series presenting the results of scientific meetings supported under the NATO Science Programme.

The Series is published by IOS Press, Amsterdam, and Kluwer Academic Publishers in conjunction with the NATO Scientific Affairs Division

Sub-Series

I. Life and Behavioural Sciences	IOS Press
II. Mathematics, Physics and Chemistry	Kluwer Academic Publishers
III. Computer and Systems Science	IOS Press
IV. Earth and Environmental Sciences	Kluwer Academic Publishers

The NATO Science Series continues the series of books published formerly as the NATO ASI Series.

The NATO Science Programme offers support for collaboration in civil science between scientists of countries of the Euro-Atlantic Partnership Council. The types of scientific meeting generally supported are "Advanced Study Institutes" and "Advanced Research Workshops", and the NATO Science Series collects together the results of these meetings. The meetings are co-organized bij scientists from NATO countries and scientists from NATO's Partner countries – countries of the CIS and Central and Eastern Europe.

Advanced Study Institutes are high-level tutorial courses offering in-depth study of latest advances in a field.
Advanced Research Workshops are expert meetings aimed at critical assessment of a field, and identification of directions for future action.

As a consequence of the restructuring of the NATO Science Programme in 1999, the NATO Science Series was re-organized to the four sub-series noted above. Please consult the following web sites for information on previous volumes published in the Series.

http://www.nato.int/science
http://www.wkap.nl
http://www.iospress.nl
http://www.wtv-books.de/nato_pco.htm

Series IV: Earth and Environmental Series – Vol. 2

Modelling of Environmental Chemical Exposure and Risk

edited by

Jan B.H.J. Linders
RIVM, Bilthoven, The Netherlands

Kluwer Academic Publishers

Dordrecht / Boston / London

Published in cooperation with NATO Scientific Affairs Division

Proceedings of the NATO Advanced Research Workshop on
Modelling of Environmental Chemical Exposure and Risk
Sofia, Bulgaria
5–9 October 1999

A C.I.P. Catalogue record for this book is available from the Library of Congress.

ISBN 0-7923-6775-8

Published by Kluwer Academic Publishers,
P.O. Box 17, 3300 AA Dordrecht, The Netherlands.

Sold and distributed in North, Central and South America
by Kluwer Academic Publishers,
101 Philip Drive, Norwell, MA 02061, U.S.A.

In all other countries, sold and distributed
by Kluwer Academic Publishers,
P.O. Box 322, 3300 AH Dordrecht, The Netherlands.

Printed on acid-free paper

Printed in the Netherlands.

CONTENTS

Foreword ix

Acknowledgements xiii

LECTURES

Pesticide Fate Models and their use. FOCUS Activities 1
Jan B.H.J. Linders

Models used in the USA for the Evaluation of Pesticide
Exposure, Hazard and Risk Assessment 17
Mark H. Russell

Quality assurance in environmental modelling 31
Gyula Dura and Elisabeth Laszlo

Possible approaches for pesticides environmental impact management 39
M.G. Prodanchuk and Alexandr P. Kravchuk

Pesticide leaching modelling validation
A Recent European Experience 47
Marnik Vanclooster

Evaporation of pure liquids from open surfaces 61
Fredric C. Arnold and Alfred J. Engel

Application of USES for estimation of PEC of pesticides
and hazard assessment for aquatic environment 73
Veska Kambourova and Kosta Vassilev

Modelling of operator exposure 79
Antonella Fait and Bengt S. Iversen

Use of alarm model in accidental pollution of Danube River
Case study. 91
Silvia Chitimiea and Aurel Varduca

Results of the use of two environmental models
for pesticides ranking by hazard 97
Fina Kaloyanova, Gyula Dura and Veska Kambourova

Recent developments in environmental modelling at Trent University, Canada ... 105
Ian Cousins, Matt MacLeod, Eva Webster and Don Mackay

MACRO: a preferential flow model to simulate pesticide leaching and movement to drains ... 117
Sabine Beulke, Colin D. Brown and Nicholas J. Jarvis

A scientific and technological framework for evaluating comparative risk in ecological risk assessments ... 133
John M. Johnston

Comparing two alternative pollutant dispersion models and actual data within an environmental health information Processing System (EHIPS) ... 151
Boris Balter, M. Stal'naya, and Victor Egorov

RBCA Toolkit: Comprehensive Risk-based modelling system for soil and groundwater clean-up ... 165
John A. Connor, P.E. Richard, L. Bowers and Thomas E. McHugh

Danish EPA use of models for assessment of pesticides mobility ... 183
Christian Deibjerg Hansen

An optimization model for the control of regional air quality in Europe ... 193
Markus Amann, Chris Heyes, Marek Makowski and Wolfgang Schöpp

Spatial refinement of regional exposure assessment ... 205
Volker Berding, Frank Koormann, Stefan Schwartz, Jan-Oliver Wagner and Michael Matthies

COUNTRY REPORTS

The future of the environmental modeling in risk assessment in Slovenia *Country Report* ... 223
Boris Kolar

Environmental Exposure of Plant Protection Products
Portuguese Experience 227
Flávia Alfarroba

Environmental Cleanup Program in Hungary
Hungarian Report 231
Eva Deseo

Modelling Experiences in the Slovak Republic
Country Report 241
Martin Murín

Dojransko Ezero (Dojran Lake) Program 1999-2001 243
Vladimir Kendrovski

Some Methodological Aspects of Soil Data Receiving
And Use for the Environmental Prognosis
Country Report 245
Galina V. Motuzova

Risk Assessment and Risk Management of Industrial
Chemicals in Poland
Country Report 251
Jan A. Krajewski

Modelling of Environmental Chemical Pollution and
Perspectives of Exposure and Risk Assessment
In Ukraine
National report 255
Leonard Dobrovolsky

Overview on Environmental Situation in Albania and
Some Issues in the Field of "Modelling"
Albanian experience 261
Tanja Floqi, Qecamedin Kodra, Genc Luarasi and Bujar Reme

CONCLUSIONS AND RECOMMENDATIONS 271

FOREWORD

Lesson one:
A model always is a simplification of reality.

More and more mathematical models are used to esimate the concentrations of different substances in the environment. An estimation of the concentrations is needed from a governmental point of view with respect to the questions whether or not to register a substance as a pesticide or to allow a substance on the market. The established estimation are used in the risk assessment procedures as part of the risk quotient. The risk quotient may be determined as a PEC over NEC ratio or as a TER, where PEC stands for Predicted Environmental Concentration and NEC for No Effect Concentration and TER for toxicity - exposure ratio. Generally the modelling concerns the estimation of the PEC in different compartments of the environment, e.g. water, soil, air, biota.

Especially in the European Union and the United States of America and Canada numerous examples of models to determine the PEC are available nowadays. The NATO Advanced Research Workshop on Modelling of Environmental Chemical Exposure and Risk was organised around four main topics:

- Outline of the characteristics of the models;
- Overview of the application of the models concerned;
- Comparison of estimated concentrations with the measured concentrations in the field and
- Credibility of modelling.

According to current guidelines in different parts of the globe exposure assessment has to be carried out for all kinds of substances, pesticides, including agricultural and non-agricultural pesticides, new and existing chemicals not being pesticides, soil pollutants, accidental pollution, etc. The participants to the workshop stressed that especially the purpose of the model should be the driving force when interpreting the results of the model calculations. Model usages should not be taken beyond their purpose. In addition it was pointed out that models always are an abstraction of reality, because of simplifying assumptions being made to keep the models within calculable limitations of scientific knowledge or computer performance.

Several types of models and modelling systems were presented during the workshop:

- FOCUS-activities of the European Union, directorate-general Health and Consumer Protection, concerning the determination of PECs in different environmental compartments like soil, groundwater and surface water; the models included here are e.g. PRZM, MACRO, TOXSWA, paper presented by Linders;
- Exposure models used in the USA, particularly at the Environmental Protection Agency, paper by Russell, including screening models like SCI-GROW and GENEEC, but also more sophisticated models as there are PRZM, TIGEM and EXAMS;
- Modelling systems presented included EUSES 1.0, a risk assessment system in the EU for new and existing chemicals, paper by Berding; USES 2.0, incorporating EUSES 1.0 and the Netherlands' evaluating system for pesticides, paper by Kam-

bourova; more toolkit modelling systems as e.g. the multimedia fugacity models, paper by Cousins, and the RCBA toolkit, paper by McHugh;

- Specific models were dealt with in more detail like the MACRO-model used for estimating the concentration in groundwater or drainage water in cracked soils (heterogeneous flow), papers by Beulke and Hansen;
- In addition, models for environmental health were presented like the EHIPS in Russia, paper by Balter, and also operator exposure calculations using the EUROPOEM databases, paper by Fait;
- Finally, the possibilities of Geographic Information Systems (GIS) were explored in several model applications, e.g. in the USA and Russia as well as in Italy and Germany.

Other items involved the quality of data and models, paper by Dura, the validation of models, paper by Vanclooster, the relation to human health perspectives, paper by Kravchuk, and several case studies, papers by Arnold, Chitimiea and Kalyoanova, presented by Dura. A specific framework for ecological modelling was presented by Johnston, while the compartment air was presented by Amann.

In the breakout sessions working groups discussed several items related to model use and development as well as co-operative actions on a regional basis. The subjects of the discussion are listed below:

- Model advantages and limitations;
- Comparative assessment of models and their use in represented regions;
- Recommendations for the use of methods for specific environmental compartments and conditions;
- Recommendations for future collaboration in model validation;
- Recommendations on future research needs on modelling.

Some of the main recommendations may be formulated as follows, while a more extensive overview of the recommendations is presented in the chapter Conclusions and Recommendations.

With respect to validation two different types of model validation were identified: internal and external validation. External validation is defined here as the comparison of model predictions with measured values. It is important that the measured data used for the validation are properly scrutinised. The spatial and temporal resolution of the data must be considered as well as the data quality.

To determine the relative importance of individual model input parameters and the uncertainty of model predictions; detailed sensitivity and uncertainty analyses should be undertaken as part of the model validation.

It is recommended that several databases should be compiled to aid the process of model validation. To avoid the duplication of measurements and to facilitate the rapid collection of measured data, it is recommended that a database containing measured concentrations is compiled.

To improve the validation status of models and the knowledge on validation and validation activities, it is recommended that international modelling efforts should be better co-ordinated. It is recommended that expert groups be established to undertake internal and external validation of existing models and to recommend future model improvements. An example of such an expert group is the "FOCUS-group" set up to investigate the large number of available pesticide leaching and surface water models.

Considering future research needs on modelling it is recommended to investigate the tools required for analysis of uncertainty. One of the key aspects of performing risk assessments is the quantification of the chance for a specific situation to occur in reality. It was considered of major importance to develop this area in the near future.

It was also recommended to develop protocols for carrying out model validation, especially how to carry out a model validation. Other areas of future development were considered to be the possibility of simultaneous exposure to mixtures of substances, the site specific scenarios in case were lower tier risk assessments indicate the potential of risk. The application of probabilistic risk assessment methods needs further development, because the aspect of occurrence of risk seems to be clearer to the risk manager. In addition a comparison should be carried out between probabilistic modelling and risk assessments making use of assessment factors.

Finally, it was strongly recommended to use all possibilities to exchange information between modellers and risk assessors, especially the information gap between east and west was indicated clearly. Training and a specific website to exchange software were considered useful. It was also recommended to organise an additional workshop on these subjects to give scientists and governmental agencies the possibility to catch-up with other, more experienced risk assessors.

ACKNOWLEDGEMENTS

The local organisation was in the good hands of Professor Fina Kaloyanova and her staff of the National Centre of Hygiene, Medical Ecology and Nutrition, which institute also hosted the workshop from 5 – 9 October 1999 in Sofia, Bulgaria. She did a very good job in making all the necessary arrangements required for participants gathering in Sofia from different parts of the world. She organised trips and took care that the participants felt themselves very well.

The editor is also grateful to the personnel of Kluwer Academic Press and in the person of Mrs. Bruins there was a fruitful helpdesk available.

Last but not least I have to thank the Scientific and Environmental Affairs Division of the North Atlantic Treaty Organisation for the approval of the organisation of the workshop and the support I got especially on the often difficult financial regulations. Always there was an ear to listen to my questions and clear enough suggestions how to proceed solved my problems. The organisers of the workshop are convinsed on the succesfulness of the event and hope that the publication of the proceedings will find its way to the many scientists, regulators and other interesting parties that could profit from it.

PESTICIDE FATE MODELS AND THEIR USE. FOCUS ACTIVITIES.

JAN B.H.J. LINDERS
National Institute for Public Health and the Environment
Centre for Substances and Risk Assessment
PO Box 1, A van Leeuwenhoeklaan 9, NL-3720, Bilthoven
The Netherlands

Abstract

Some years ago the Commission of the European Union established FOCUS (FOrum for the Co-ordination of pesticide fate models and their USe). Several working groups made inventories of the characteristics of fate models for leaching to groundwater, soil and surface waters. Based on the recommendations new working groups were formed on leaching/soil scenario development and surface water scenario development. The results and the recommendations of the models to be used together with the scenario development process are described. In addition, the use of the scenarios in the registration procedure to place plant protection products on the market according to Directive 91/414/EEC is elucidated.

1. Introduction

In 1994 the European Commission established the FOCUS organisation. FOCUS stands for FOrum for the Co-ordination of pesticide fate models and their USe. The European Directive 91/414/EEC deals with the process for placing plant protection products on the market. In accompanying Annexes the data requirements (Annex II for the active substances and Annex III for the products) and the final Uniform Principles (Annex VI, 97/57/EEC) are mentioned. The Uniform Principles describe the criteria that must be fulfilled by the products to be put on the market in the Member States and also the way the registration authorities have to determine the concentrations to be expected in the environment after the application of the products. It is stated that "a suitable and at Community level validated calculation model" should be used for the environmental compartments soil, groundwater and surface water. For this reason three FOCUS Working Groups were established:

1) leaching to ground water,

J.B.H.J. Linders (ed.), Modelling of Environmental Chemical Exposure and Risk, 1–15.

2) surface water and
3) soil.

All three groups have given an overview of the existing and useful models at the area of concern. Three reports (DOCs 4952/VI/95 [1], 6476/VI/96 [2] and 7617/VI/96) [3] were adopted as guidance documents by the Standing Committee on Plant Health (SCPH) in Brussels to be used by all the Member States in performing risk assessments for registration purposes of plant protection products. As a follow-up of this activity and also based on the conclusions of the different groups the FOCUS Steering Group thought it useful to establish two new Working Groups on the development of European scenarios, one for groundwater/soil scenarios and one for surface water scenarios. The current paper describes the process the different Working Groups have followed to select the relevant models and to develop these scenarios. The underlying reasoning for the choices and some preliminary conclusions and recommendations are mentioned. The emphasis will be put on the compartment surface water. The work on leaching and soil modelling and scenario development will be dealt with shortly.

2. FOCUS - Organisation

The Council Directive 91/414/EEC of 15 July 1991 concerning the placing of plant protection products on the market describes the requirements which have to be fulfilled in order to obtain an authorisation for a plant protection product.

The Directive has given great importance to the calculation of Predicted Environmental Concentrations (PEC) which are then used for conducting further experiments or as a support for evaluation and decision making. It is suggested that PECs are calculated using a suitable model or calculation method.

Since the regulatory use of simulation models is quite new, there are presently neither clear and detailed guidelines nor a generally agreed practice on how they are to be used and how the results are to be interpreted. The role and importance of models for calculation of PECs depend strongly on their quality, which can be established through a validation process. Currently no models are validated at a community level.

The intention of FOCUS is to provide industry and regulators with expert advise on the state-of-the-art in simulation modelling and to give the research community a clear view of deficiencies in the present state of simulation models when they are to be used in a regulatory context. FOCUS is a group of regulators, industry representatives and experts from research institutes. The work is co-ordinated by a Steering Committee. The current organisation is depicted in Figure 1.

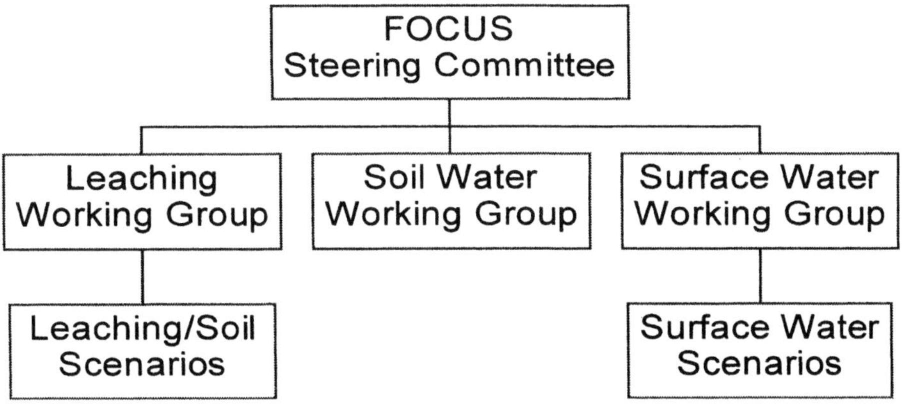

Figure 1. Organisation of FOCUS

The three Working Groups on Leaching models, Surface Water models and Soil models have prepared guidance documents that were released by the European Commission. After the adoption of these documents and the official acceptation as guidance documents the Steering Committee decided to install two follow-up groups for the development of European scenarios, one for leaching/soil and one for surface water. Both latter groups are currently in the finalising stage of report production. The current status and progress is also reported. As an example the work of the working group on surface water is treated in more detail.

3. Surface Water Entry Routes

The main entrance routes of plant protection products and active ingredients into surface waters were identified as:

- Spray drift
- Surface run-off
- Drainage
- Atmospheric Deposition.

Of course there are other routes possible e.g. incidental releases like cleaning tanks, or accidental releases, but these were considered not in conformity with Good Agricultural Practice (GAP) and were thus considered for inclusion in the modelling process for registration. Atmospheric deposition was not considered by the group to be a major entry route and because work is already carried out on this topic by other fora such as the European and Mediterranean Plant Protection Organisation (EPPO). For the other entry routes available models will be dealt with in the following sections.

4. Describing Fate in Water

The mathematical description of fate and behaviour of pesticides in the environment is always based on common scientific principles such as conservation of mass. It should be born in mind that a model is a simplification of reality. The simpler the model the greater is the deviation from real observed phenomena. On the other hand, whereas a complex model may simulate many actual processes, it will require a far wider range of input data describing environmental parameters, which may be difficult and/or expensive to measure. The models described in this paper are considered to be the most sophisticated models currently available depending on the intended use. Factors determining the usability of a model are related to the following items:

- Load. Which discharges into the surface water body are taken into account.
- Mathematics: What mathematical representations are used for the processes taking place, e.g. linear or Freundlich sorption, first order or Michaelis-Menten kinetics for the degradation.
- Solutions: Which solutions are possible for the differential equations: analytical or numerical.
- Validation: Is the model validated using independent data, and have a sensitivity analysis and a model verification process been carried out.
- Sediment: The possibility of calculating the PEC in sediment has been added to the remit of the group and this item has therefore been included in the evaluation process.

5. The Models

After cataloguing the available models that are currently, or could be used in a regulatory context; those presented in Table 1 were selected for a more in depth description.

TABLE 1. Selected models for the several items of consideration

Drift	Drainage	Run-off	Atmospheric Deposition	Fate
IDEFICS	CHAIN_2D 1.1	EPIC	none	ABIWAS
MOPED	CRACK_P 1.0	GLEAMS		EXAMS
PEDRIMO	MACRO 3.1	OPUS		SLOOT.BOX
PSMDRIFT	OPUS 1.63	PELMO		TOXSWA
TABLES	PESTLA 3.0	PRZM2		WASP
- NL	PESTRAS 2.1	SWRRBWQ		
- UK				

The models have been arranged in alphabetical order.

In the final report of the group these models are compared with each other for an extensive list of items, including:

1. General Information
2. Documentation and systems considerations
3. Model Science

The items may vary depending on the entry route considered. The example presented refers to the fate models in particular.

The models finally selected for use in the estimations of the groundwater concentration were PELMO, PRZM, PESTLA and MACRO. These models are not dealt with here in detail. The PESTLA-model will be replace by the PEARL-model in the near future.

6. Advantages

Each model has its own specific advantages and disadvantages. These include how simple or complex the model is and also, the purpose for which it was developed. For example, a model designed principally to simulate a particular process such as surface run-off, may perform better in predicting surface run-off inputs, than another, more general model designed to simulate a range of processes including leaching drainage, but which simulates surface run-off in a less mechanistic way. This may seem trivial but should always be kept in mind when evaluating models.

The group considered the following criteria to be of major importance when assessing model advantages:

- PEC in sediment; the potency of the model to estimate a concentration in the sediment phase of the aqueous environment was part of the remit of the group and is part of the data requirements of guideline 91/414/EEC. On the other hand, models not considering the sediment phase estimate the concentration in the aqueous phase and from that estimate a concentration in sediment can be estimated using equilibrium partitioning.
- Ease of use; because the models considered are to be used in a regulatory context - and because in such a context, model users are unlikely to be the model developers or researchers, their user friendliness is important. The availability of standard scenarios for the regions under concern still have to be developed.
- Commonly available; when a model is used in regulatory decision making for the registration of plant protection products in the EU it should be easily available to all potential users.
- Validated; the guideline 91/414/EEC states that if the possibility exists the concentration should be estimated using a suitable and on community level validated model.

6.1 SPRAY DRIFT

When considering the possibilities for including spray drift as a contamination route for surface water PEC calculations there are two possible approaches:

1. tables or fixed values for different applications and

2. models simulating the droplet distribution in pesticide sprays during application.

The first approach has the advantage of simplicity, although this can be taken to far. The UK, for instance, assumes a worst case with the process of overspray: the ditch itself is sprayed with the same amount of pesticide as the field. Being non-GAP, however, it certainly serves as a safety estimate. The Dutch and German tables (PSMDRIFT is based on a tabular approach, although it can interpolate a drift factor for every distance to the ditch), give a range of estimated 'drift factors' at the edge of the field, the application method and the target crop scenario.

The second approach is the most promising for the future. They are based on recently carried out research and take into account different application parameters such as meteorological characteristics (wind speed, wind velocity) and apparatus characteristics (type of nozzle, nozzle direction, boom height, spray volume, droplet distribution).

6.2 DRAINAGE

None of the models evaluated were considered to be the first among equals. Whereas specific models considered to be superior in simulating some aspects of the drainage entry route, none were considered to be superior in all aspects. OPUS combines surface runoff and drainage, CHAIN_2D is the only two-dimensional model, PESTLA and PESTRAS contain the most advanced and flexible descriptions of chemical and biological processes, MACRO and CRACK_P can account for preferential flow. Therefore, a very powerful tool would be a model that combines the positive features of the models mentioned.

6.3 SURFACE RUN-OFF

All the surface run-off models considered are based on the same principles: curve numbers and the (modified) universal soil loss equation. They do however, vary in associated processes considered, e.g. PRZM2 and PELMO can be chosen if volatilisation is an important factor, OPUS or GLEAMS allow simulation of non-homogeneous slopes, while EPIC appears to be very useful in simulating several agronomic scenarios. Although SWRRBWQ is the only model to include multiple field flows within a basin, it has no other advantages over any of the other models considered.

6.4 FATE IN SURFACE WATER

The models that actually calculate the PEC for surface waters can be separated into two categories: simple, screening models and complex, sophisticated, level 2 or 3 models. The two simple models are ABIWAS and SLOOT.BOX, the complex models are EXAMS, TOXSWA and WASP. Again the models vary in their specific features.

ABIWAS uses abiotic degradation rate constants, whereas SLOOT.BOX was developed primarily as a registration tool to calculate the concentration in surface water by treating all disappearance routes as a lumped degradation rate constant. The models EXAMS and WASP can also be used in branched systems and in addition they take into account metabolites, while TOXSWA also accounts for sorption onto macrophytes.

7. Disadvantages

The disadvantages and/or limitations of the models presented are more or less the opposite of the advantages. More specifically the following items are taken into account:

- Limited to Water Phase; several surface water fate models do not calculate the concentrations of a pesticide in the sediment phase. Therefore, another method is needed to take care of this calculation. Mostly, equilibrium partitioning between the water and sediment phases is used as an approximation of the distribution. If the sorption capacity to soil and/or sediment is known a reasonable estimation is possible, otherwise the n-octanol/water partitioning coefficient can be used to make an estimation.
- Complex Expert Use; most of the models have just recently been developed or are still under development. The experience with the models is therefore generally limited to the researcher or developer of the model. Because of this, most regulatory users are not familiar with the models, their limitations and what may be the most suitable standard scenarios to use.
- Research Tool; in relation to the former point it is clear that the models are often used as research tools in the hands of the developer. Thorough testing by the researcher should be normal practice, of course, but once developed, specific research versions need to be adapted, calibrated and validated for a specific regulatory usage. If further model development by the researcher then takes place, this process has to be repeated before the updated model can be used for regulatory purposes.
- Invalidated; the current validation status of all the models is considered to be low. Some models are only validated in a very specific situation. The only model for which a systematic validation is underway is TOXSWA. A lot of work is needed before any model can be considered to be validated at the community level.

7.1 SPRAY DRIFT

The limitations of the drift tables are clearly the fixed values at which the drift factor has been set. However, for screening purposes and for ranking pesticides they can be very useful. The drift models, having been developed very recently, lack validation and can be considered still as research tools. The intention to validate these models, although very much based on field research, should be strongly supported. None of the models IDEFICS, MOPED nor PEDRIMO take into account specific pesticide data in estimating the drift factor during application. The model IDEFICS is limited to boom sprayers, while MOPED considers only horizontal spraying.

7.2 DRAINAGE

In table 2 an overview is given of the main limitations or deficiencies of the drainage models presented. It shows that several models have the same limitations, which can be due to the fact that it can be quite difficult to develop an accurate mathematical description of the process.

TABLE 2. Drainage model

Model	Main limitations/deficiencies
CHAIN_2D 1.1	Degradation not affected by water content
	Difficult to use for routine applications
	Non-structured soils only
CRACK_P 1.0	Clay soils only
	Outputs sensitive to (uncertain) macropore-related parameters
MACRO 3.2	Outputs sensitive to (uncertain) macropore-related parameters
OPUS 1.63	One-dimensional treatment of solute flux in saturated zone
	Incomplete documentation
	Non-structured soils only
PESTLA 3.0	Non-structured soils only
PESTRAS 2.1	No field drains
	Non-structured soils only

7.3 SURFACE RUN-OFF

Most of the run-off models suffer from the same limitations:

- with the exception of PELMO, all models assume equilibrium partitioning between pesticide sorbed to the soil surface and in the runoff water. This results in an overestimated concentration in runoff water.
- curve number approach is derived from empirical studies in the USA and is, therefore not validated for either the EU or other parts of the world.
- all of the models work on a daily time step, whereas actual run-off events can occur on an hourly time frame. On the other hand if hourly time-step models were to be developed, then they may suffer from a lack of available rainfall input data.
- only uniform slopes and land use can be simulated, except for OPUS and GLEAMS.

7.4 FATE IN SURFACE WATER

The limitations of the screening models are obvious: they are only screening models and therefore suitable for the estimation of PECs in surface water at a simple screening level. When such estimations are required the models, ABIWAS and SLOOT.BOX may serve their purposes quite sufficiently. ABIWAS estimates only the abiotic degradation of substances while SLOOT.BOX also takes into account biodegradation and physical disappearance, advection, sorption, sedimentation and resuspension.

When considering more complex, mechanistic models the following limitations can be mentioned:

- all models describe the water flow as a steady state. When longer periods have to be examined this assumption may be less realistic.
- all models assume instantaneous mixing over the cross-section of the segments. TOXSWA does not allow for a definition of vertically stacked segments while WASP and EXAMS do. It was shown that it could take about 24 hours before the pesticide entering the water body has been spread to the lower parts of the watercourse.
- estimation of PEC_{sed} may be influenced by the schematisation of the sediment phase necessary for the discretisation of the differential equations.
- WASP and EXAMS describe sorption as an instantaneous and linear process, while TOXSWA also allows a Freundlich type of adsorption.
- TOXSWA uses a single lumped degradation coefficient, while EXAMS and WASP describe the different degradation processes separately and therefore allow the possibility of correcting for light intensity, pH or temperature.
- WASP and EXAMS do not have the possibility to take into account sorption onto macrophytes; TOXSWA does.

8. General Approach for Scenario Selection

To develop a scenario for a calculation method for a concentration of an active substance in surface water several items are relevant. In the dossier package a registrating company (or registrant) has to deliver to the governmental authorities, there are a lot of data on compound specific information relevant for the environmental compartments of the ecosystem, like:

- application data, i.e. dosage, frequency, interval;
- physico-chemical data, i.e. melting and boiling point solubility, vapour pressure, octanol-water partitioning coefficient K_{ow}, dissociation constant pKa;
- fate data, i.e. degradation in water/sediment systems, degradation in soil, hydrolysis, photolysis, sorption characteristics;
- ecotoxicological data, i.e. toxicity for water organisms, for birds, for earthworms, for micro- and macro-organisms, etc.

Some of these data is needed for the calculation of the Predicted Environmental Concentration (PEC) in surface water. There are, however, also other data a model may need before it is used for this purpose, the scenario data. To this type of data belong, e.g. crop, agronomic parameters like tillage, environmental parameters like hydrogeology, meteorology and soil data.

The following definitions are used in the context of development of EU-scenarios:

- <u>scenario</u>: a representative combination of crop, soil, climate and agronomic parameters to be used in modelling; representative means that the selected scenario should represent physical sites known to exist, i.e. the combination crop, soil, climate and agronomic conditions should be realistic;

- scenario data: freely chooseable information required to run the model applied in a specific situation and related to agricultural (crop, agronomic parameters like tillage), environmental (hydrogeology, surface water characteristics), climatic (meteorology), and soil (pH, % organic matter) characteristics.

The scenario definition calls for information on crops, soils, climate (precipitation and temperature), land use and steepness data. Several databases were consulted to find the relevant data to determine the areas in Europe that could be examples for the intended purposes. Table 3 shows the final selection and nomenclature of the scenarios to be used in the calculation models as determined by the working group.

In addition to the data summarised in Table 3, data on crops and types of surface waters (water bodies) are needed. In identifying the most relevant crops in Europe a list of crops was prepared taking into account the importance of the crop in aereal terms and distribution over Europe. The following crops were selected: cereals (not maize), maize, potatoes, sugar beet, oil seed rape, sunflower, soybean, tobacco, hops, vegetables, pome/stone fruit, citrus, vines and olives. For these crops specific data needed to be collected, some being scenario independent, like leaf area index, etc., others being scenario dependent, like emergence time, harvest date, etc. Using 1:25.000 maps of the areas a choice was made on the water bodies present in the locations, like ditches, small streams or ponds. Also the dimensions of the water bodies were determined. In Table 4 an overview is given of the different water bodies. It is assumed that the dimensions do not differ for the locations.

Figure 3. gives a representation of the distribution of soil/leaching and surface water locations selected for scenario development in Europe.

9. Weather Data

The weather data of the selected locations were analysed statistically for the mean weather year in the time period available. The MARS-project of the EU Joint Research Centre in Ispra, Italy, provided the meteorological data used. An additional requirement for the selection of the mean year was the presence of relevant storm events during the spring period especially for the run-off locations. To be able to run all the models a period of 16-month starting with the mean year was used to perform the calculations.

10. Water Bodies

Three water types have been selected for the application of the scenarios: a small ditch, a stream and a pond. The determined default values for these water bodies are given in table 4.

The indicated values are not intended to be definitive but should be seen as an expert judgement's view on the intended reasonably worst-case situation.

11. Final Model Choice

The inputs to water bodies after application of plant protection products are drift, drainage, run-off and atmospheric deposition. Atmospheric deposition is still not taken into account because of missing mathematical instrumentation. Drift will be described by interpolating the drift data as presented by Ganselmeier *et al.* [4], drainage is calculated by the model MACRO [5], run-off by the model PRZM [6] and finally, the fate of the substance in surface water by the model TOXSWA [7]. The way the models are connected to each other in the calculation sequence is given in Figure 2.

Table 3 Defined scenarios

Code	Climate	Soil	Temperature	Precipitation	Slope	Type	Weather station
D1	Scandinavia	Clay	Cold	Moderate	Gentle	Drainage	Lanna
D2	North-west Europe	Clay	Temperate	Moderate	Gentle	Drainage	Brimstone
D3	Northern maritime	Sand	Temperate	Moderate	Flat	Drainage	Vredepeel
D4	Northern maritime	Loam	Temperate	Moderate	Gentle	Drainage	Skousbo
D5	Western maritime	Heavy loam	Temperate	Wet	Moderate	Drainage	La Jaillière
D6	Eastern Mediterranean	Heavy loam	Warm	Moderate	Gentle	Drainage	Thebes
R1	Middle European land	Silty	Temperate	Wet	Gentle	Run-off	Weiherbach
R2	Atlantic southern maritime	Loamy	Temperate	Very wet	Very steep	Run-off	Porto
R3	Middle European Mediterranean	Sandy loam	Warm	Wet	Steep	Run-off	Bologna
R4	Southern European Mediterranean	Loamy	Warm	Moderate	Moderate	Run-off	Roujan

Table 4: Parameterisation for water bodies.

Variable	Ditch	Stream	Pond
Depth (m)	0.3	0.5	1.0
Replacement time (days)	50	0.1	50
Distance from field to water's edge (m)	0.5	0.5	3.0

12. Current Work

The working group will present the final report in spring of the year 2000, including the description of the work done, the final input data for the different scenarios, using some example data of existing active substances and the results of the calculations following the sequence of figure 2. Input files are being prepared for crops, weather, soil and other parameterisations required for the models, like dimensions of the area under consideration, the amount of surface water, the dimensions of the water bodies, etc.

The scenarios are intended to be used in the EU for the registration of plant protection products. It is of vital importance that the approaches proposed are understood and agreed in the Member states and that there is a willingness to use the models and the scenarios. Therefore, an intensive training and familiarisation programme will be started to introduce the scenarios in the Member States and industry. It is the intention to prepare a CD-ROM containing the models, the scenarios and the necessary documentation.

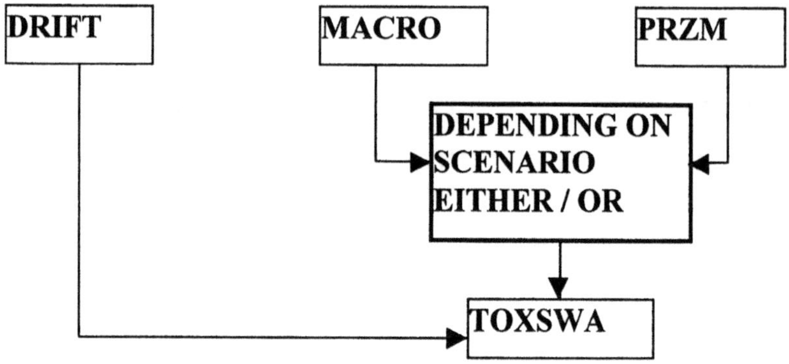

Figure 2. Calculation sequence of models.

The final report will also contain information on the Geographical User Interfaces of the different models to help the user in getting started with the models in the process of the evaluation of data concerning the determination of the Predicted Environmental Concentration (PECs) as described in Directive 91/414/EEC.

13. Conclusions and Recommendations

The main conclusions of the FOCUS Surface water Modelling Group are:
There are currently no at the European level validated models available. Several models are locally validated or being validated to a limited extent. Other models are not validated at all and because they are not further developed will not be validated in future.
For some models the validation status is rising. Validation studies are in progress already or have been scheduled in the near future. These models must be considered as the most promising ones.
There are currently several useful models available for simulating surface water loadings via the various entry routes defined by the group. For spray drift, the best results are obtained using drift tables, e.g. the German drift table combined with the simple interpolation model PSMDRIFT. With respect to drainage the most useful models are considered to be PESTLA, CRACK_P and MACRO. Concerning runoff, the models GLEAMS, PELMO and PRZM are considered to be most applicable. And finally, for fate the models EXAMS, TOXSWA and WASP are considered to give the

most useful results in estimating the concentrations of pesticides in surface waters and sediment.
During the evaluation of surface water behaviour of pesticides a tiered approach must be considered most promising, because the most detailed and complicated modelling is only required when absolutely necessary. In particular, the screening models ABIWAS and SLOOT.BOX may give useful first or second step results in combination with a defined standard European scenario.
As the European evaluation of pesticides is just starting it is not surprising that standard European scenarios are lacking. However, this is considered to be a serious problem for the development of a harmonised European approach to estimating PECs.

Finally, the group comes to the following main recommendations:
Research should be carried out for drift data in Southern Europe. All of the drift data come from countries in the west or north of the European Union. It is questionable if these data can be extrapolated to southern Europe. However, efforts should be made to extrapolate and validate the current models for southern European conditions.
Validation of all models considered is urgently needed, especially in view of the wording in the EU guideline 91/414/EEC. If validation at community level is not yet possible, models should only be used for the situation they are validated. In particular, validation efforts should be focused on the following:

- Runoff curve numbers, as they are only empirically established for US situations,
- Drainage, especially on the community level,
- Fate, as work in this area has been started only recently, certainly with respect to the model TOXSWA.

Development of European scenarios. Registration of active ingredients has to be approved by the Commission of the EU taking into account European circumstances. Only the registration of specific products belongs to the competence of the local designated national authority. This common registration procedure can not function without the availability of suitable scenarios within the European Union.
There is no model available describing all the input routes and behavioural aspects of plant protection products in the European Union. Such a model could be constructed building on elements of the available models for the different input routes and the fate models themselves. As has been shown in the example calculation using output of one model as input to the next model is possible but is not considered easy. It is time consuming and expensive. Streamlining this process is strongly recommended.

Interpretation of model results. An independent problem arising from using models is the interpretation of the model results, certainly in the light of the consequences for the registration or refusal of a registration in the EU. Model developers, model users and decision-makers should work together in gaining knowledge on how to interpret the results and if necessary to carefully balance an appraisal.

At the moment of the presentation of this paper a lot of work has been done or is carried out in the near future. Therefore, it should be clear that nothing has been finalised yet and the results obtained should be considered as draft and treated with care.

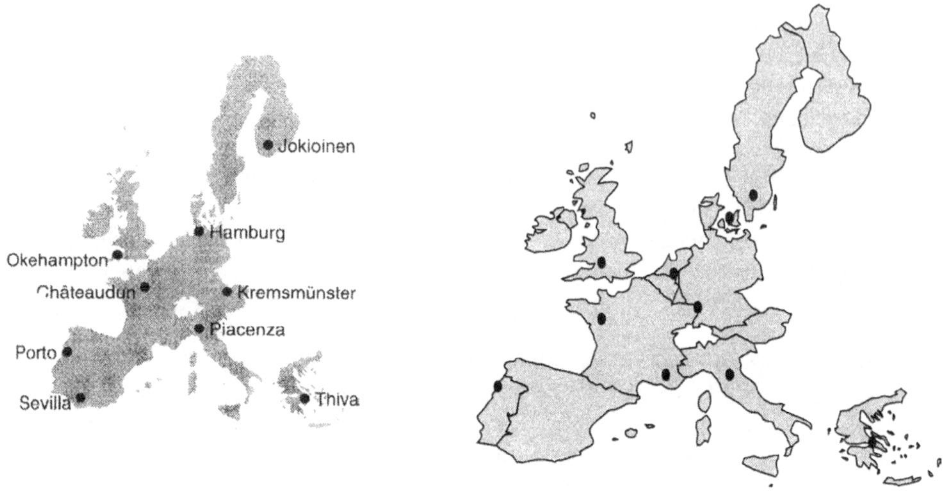

Figure 3. *Areas/locations of the European Scenarios for surface water and soil/leaching.*

Currently, ten scenarios have been developed for use in PEC calculations for surface water, intended for the decision making within the framework of 91/414/EEC. The scenarios take into account several parts of Europe, with specific properties concerning soil, weather, crops and surface water bodies. The approach taken is a stepwise or tiered method in which the results of the PEC estimation may be compared to acute and chronic toxicity data for different species of aquatic organisms. If at a low tier the relevant trigger values are exceeded the next tier comes into operation. The working group believes that a useful tool has been reached to bring the process of risk assessment of plant protection products to a higher level of sophistication in which risk assessors and risk managers of different interests can have faith. More work needs to be done on the validation of the models, although the models use the current state-of-the-science in the mathematical description of processes occurring in the environment and in this case the aquatic compartment. Experts in fate and behaviour of plant protection products and experts on ecotoxicology have to work close together to further develop the comparison of modelling results and results of ecotoxicological testing. If agreement may be reached in this comparison the risk assessment process for regulatory use in the framework of 91/414/EEC can make a great step forward.

14. References

1. Boesten, J.; Businelli, M.; Delmas, A.; Edwards, V.; Helweg, A.; Jones, R.; Klein, M.; Kloskowski, R.; Layton, R.; Marcher, S.; Schäfer, H.; Smeets, I.; Styzcen, M.; Russell, M.; Travis, K.; Walker, A. & Yon, D.; (1995) - Leaching Models and EU Registration. European Commission Document 4952/VI/95, Brussels, 123 pages.

2. Boesten, J.; Helweg, A.; Businelli, M.; Bergstrom, L.; Schäfer, H.; Delmas, A.; Kloskowski, R.; Walker, A.; Travis, K.; Smeets, L.; Jones, R.; Vanderbroeck, V.; Van der linden, A.; Broerse, S.; Klein, M.; Layton, R.; Jacobsen, O-S & Yon, D.; (1996) - Soil Persistence Models and EU Registration. European Commission Document 7617/VI/96, Brussels, 74 pages.

3. Adriaanse, P.; Allen, R.; Gouy, V.; Hollis, J.; Hosang, J.; Jarvis, N.; Jarvis, T.; Klein, M.; Layton, R.; Linders, J.; Schäfer, H.; Smeets, L. & Yon, D. (1996) - Surface Water Models and EU Registration of Plant Protection Products. European Commission Document 6476/VI/96, Brussels, 218 pages.

4. Ganselmeier *et al.* (1995), Spray drift of plant protection products, BBA, Heft 305, Berlin, Germany, 111 pp.

5. Jarvis and Larsson (1998), The Macro model (version 4.1): Technical Description. www.mv.slu.se/macro/doc/.

6. Carsel *et al.* (1998), PRZM-3, A model for predicting pesticide and nitrogen fate in the crop root and unsaturated soil zones: Users Manual for Release 3.0. U.S. Environmental Protection Agency, Athens.

7. Adriaanse (1996), Fate of pesticides in field ditches: the TOXSWA simulation model, SC-DLO Report 90, Wageningen, The Netherlands, 203 pp.

MODELS USED IN THE USA FOR THE EVALUATION OF PESTICIDE EXPOSURE, HAZARD AND RISK ASSESSMENT

MARK H. RUSSELL
DuPont Agricultural Products
Barley Mill Plaza 37-6162
Wilmington, DE 19880-0037
USA

Abstract

A wide range of models are currently being used by the United States Environmental Protection Agency (USEPA) Office of Pesticide Programs to provide regulatory estimates of environmental exposures from labelled uses of pesticides in agriculture. Additional models and modeling concepts have been developed to address various aspects of hazard assessment as well as the overall evalaution of ecological risk resulting from pesticide use. Current groundwater exposure models include SCI-GROW and PRZM3 while surface water estimates involve use of GENEEC, PRZM3, EXAMS and MUSCRAT. A number of probabilistic exposure modeling and risk assessment concepts have been endorsed by a recent EPA workgroup called ECOFRAM (Ecological Committee on FIFRA Risk Assessment Methods).

1. Introduction

Currently, a number of models are being used by the United States Environmental Protection Agency (USEPA) to provide regulatory estimates of environmental exposures from labelled uses of pesticides in agriculture. There are also several models and modelling concepts that have been developed to address various aspects of hazard assessment and the overall evaluation of ecological risk resulting from pesticide use. This paper provides an overview of the major existing models and discusses a number of modelling concepts that have been proposed for further development in the USA.

The process of ecological risk assessment (ERA) involves the assessment of environmental exposures as well as the evaluation of hazard values in order to determine the level of risk for exposed organisms. Various types of models are currently being used for the assessment of exposure, hazard and risk.

J.B.H.J. Linders (ed.), Modelling of Environmental Chemical Exposure and Risk, 17–30.

2. Exposure Assessment

The USEPA currently requires a relatively complete battery of environmental fate data to support registration of a new pesticide [1]. These data include physical/chemical information, studies on the rate of degradation due to hydrolysis, photolysis and microbial action, and evaluations of sorption and mobility in soil. Groundwater evaluations are typically performed using screening modelling which can then trigger small-scale prospective groundwater studies. Groundwater modelling is normally used to help rationalise the results of existing field studies and to minimise the need for additional fieldwork. The current tiers of exposure assessment used by the USEPA for surface water include both screening modelling as well as more detailed mechanistic modelling. The overall approach of exposure assessment relies on the results of laboratory studies, modelling assessments and field studies. (Figure 1.).
Figure 2. shows the relation between the expected range of concentrations calculated in the different tiers and the actual concentration considered possible.

2.1 GROUNDWATER EVALUTIONS

The current tiers used for the evaluation of potential impacts of pesticides on groundwater include a screening evaluation using the regression model SCI-GROW (Screening Concentration in Groundwater) [2] followed by a field study of leaching if the combined results of the laboratory testing, field dissipation studies and SCI-GROW indicate significant leaching concerns.

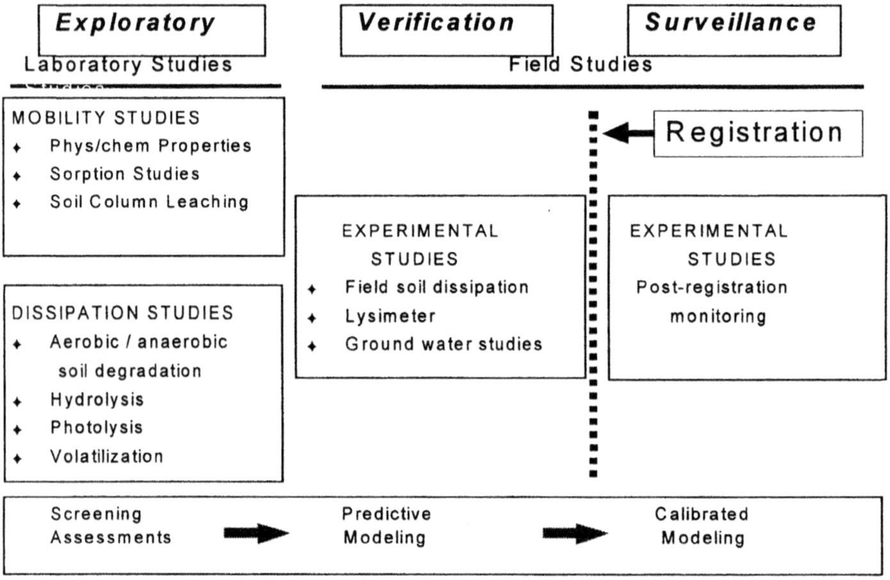

Figure 1. Progressive use of experimental and modelling data for exposure validation

A summary of the groundwater screening model SCI-GROW is given in table 1 and an example of SCI-GROW calculations in table 2.

TABLE 1. Summary of current groundwater screening model, SCI-GROW

Category	Description
Current model Version	SCI-GROW 1.0
Theoretical basis	Regression model based on the results of small-scale prospective groundwater studies in the USA
Key assumptions	Highly vulnerable settings (sandy. Low OC soils, high annual rainfall, shallow groundwater)
Key inputs and outputs	Chemical use rate, sorption coefficients, degradation in soil

TABLE 2. Example of SCI-GROW calculations

Use rate (lb ai/A)	Koc (ml/g)	Soil half-life (days)	Peak 4-month-average GW Concentration (μg/L)
1	1000	80	0.086
1	200	20	0.093
1	100	10	0.040
0.5	1000	200	0.071
0.5	200	30	0.079
0.5	100	20	0.076
0.05	1000	300	0.009
0.05	200	200	0.066
0.05	100	200	0.070

Mechanistic modelling of leaching is performed using PRZM3 [3], which incorporates chemical, climatic, soil and agronomic data to provide estimates of concentrations in soil and soil pore-water as a function of depth and time. PRZM3 results can be obtained either deterministically (i.e. a single year) or probabilistically (i.e. a summary of multiple years) to provide an evaluation of the leaching potential of a pesticide. The tables 3, 4 and 5 give an overview of the regulatory strategy in evaluating the leaching potential of pesticides. An example of the deterministic and probabilistic results of PRZM3 is given in the figures 3 and 4.

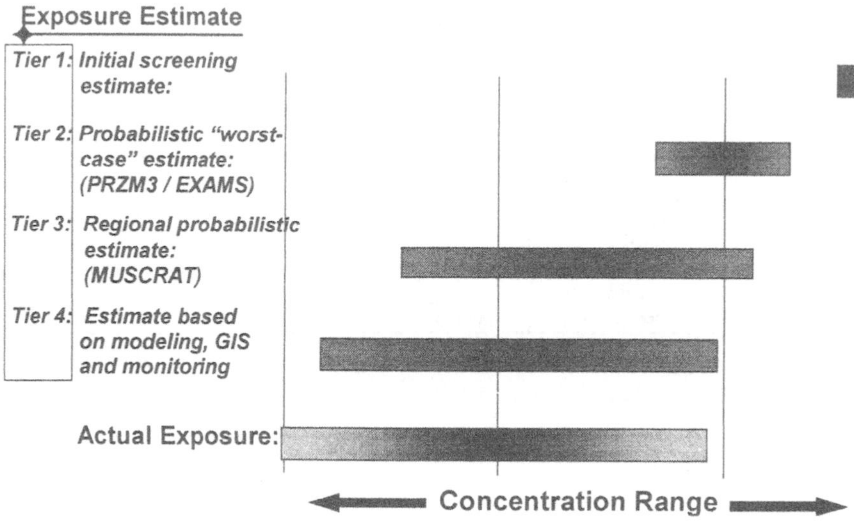

Figure 2. Basic concept behind modelling tiers

TABLE 3. Summary of current regulatory model, PRZM3

Category	Description
Current model Version	PRZM3, Version 3.12
Theoretical basis	Storage routing (tipping bucket) hydrology, convection-dispersion solute transport
Key assumptions	No upward movement of water, linear, reversible sorption, first order degradation
Key inputs and outputs	Chemical properties, soil properties, climatic data, agronomic data, chemical concentration in soil and soil-pore water

TABLE 4. Summary of input data required for deterministic leaching modelling

Type of input data	Description
Chemical	Needed for parent and metabolites (degradation kinetics, sorption, solubility, vapour pressure)
Climatic	Daily data (precipitation, air temperature, evapotranspiration)
Soil	Profile data (association, texture, OC, moisture capacities/tensions, pH and structure with depth)
Agronomic	Typical practices (chemical application rate and date, application method, crop type, soil and water management practices, field characteristics)

TABLE 5. Output data from leaching modelling

Length of simulation	Description of outputs
One year	Mass balances of chemical and water over time; concentrations in soil, groundwater and surface water
Multiple years	Mass balance of chemical and water over time; concentrations in soil, groundwater and surface water; long term concentrations trends; probabilistic estimates of concentrations

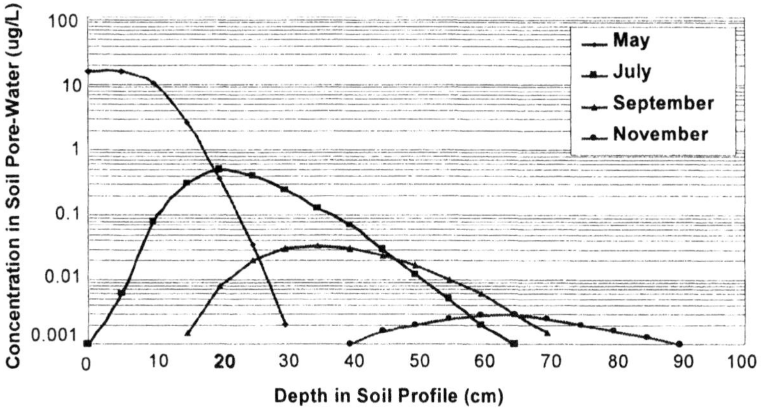

Figure 3. Example of Deterministic PRZM-3 Output

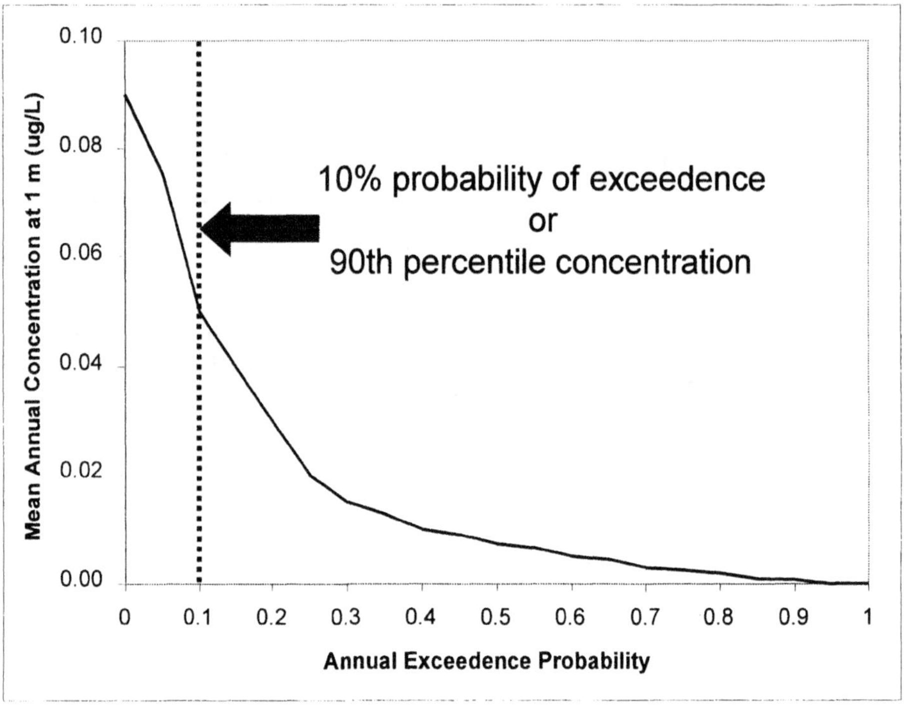

Figure 4 Example of Probabilistic Output from PRZM-3

2.2 SURFACE WATER EVALUATIONS

In table 6 an overview is given of the modelling tiers used by USEPA for evaluating the impact of plant protection products in surface waters. The current modelling tiers for surface water are performed using the meta-model GENEEC (Generic Estimated Environmental Concentration) [4]. This model estimates the potential concentration of pesticide in a static pond receiving drift, runoff and erosion from an adjacent agricultural field. GENEEC is intended to provide a highly conservative (i.e. high) initial estimate of the potential concentrations of a pesticide in surface water. Table 7 provides an overview of the screening model GENEEC, while table 8 shows an example calculation of this model. Actual surface water concentrations are likely to be lower due to the effects of actual runoff and erosion levels, buffer zones, water movement in the receiving water body and volatilisation. Tier 2 modelling of potential concentrations in surface water is performed using PRZM3 to provide estimates of runoff and erosion and EXAMS [5] to simulate a receiving water body. The combined PRZM3 / EXAMS models simulate the effects of specific combinations of soil types, crops and climatic inputs on the predicted concentrations in a farm pond.

In table 9 the current combination of the regulatory models PRZM3 / EXAMS is given with key parameters and assumptions.

TABLE 6. Current modelling tiers for surface water in the USA

Tier	Description
1	Deterministic modelling using GENEEC; worst case concentration calculated
2	Probabilistic modelling using PRZM3 and EXAMS; 1 – 3 highly vulnerable sites; 90th percentile concentration calculated
3	Probabilistic modelling using MUSCRAT; 25 scenarios per region, varying soil and precipitation; 90th percentile concentration calculated
4	Landscape modelling, GIS

TABLE 7. Summary of current screening model, GENEEC

Category	Description
Current model Version	GENEEC, Version 1.2
Theoretical basis	Meta-model based on PRZM / EXAMS
Key assumptions	10 ha of land drain to a 2 m deep, 1 ha static pond; chemical loss from field is via run-off and erosion
Key inputs and outputs	Chemical use rate, application method, sorption coefficient, degradation rate due to hydrolysis, photolysis and aerobic metabolism; time-weighted average concentrations in surface water

TABLE 8. Example of GENEEC calculations

Model input and output parameters	Value
Inputs	
Application rate	1.0 kg as/ha
Application method	Ground spray, not incorporated
Sorption coefficient	100 ml/g
Soil aerobic half-life	30 days
Aquatic aerobic half-life	30 days
Photolysis half-life	10 days
Outputs	
Time-weighted-average concentrations of chemical in static receiving waters	
0 day TWA	3.93 μg/L
4 day TWA	3.72 μg/L
21 day TWA	2.79 μg/L
28 day TWA	1.70 μg/L

TABLE 9. Overview of current regulatory surface water models: PRZM3 / EXAMS

Category	Description
Current model Version	PRZM3, Version 3.12; EXAMS 2.97.5
Theoretical basis	EXAMS – constant volume / flow; well-mixed compartments with dispersion
Key assumptions	EXAMS – linear, reversible sorption, first-order and second-order degradation via multiple mechanisms
Key inputs and outputs	Compartment properties (physical and chemical), chemical input loadings (e.g. drift, run-off and erosion); chemical concentrations in compartments over space and time

TABLE 10. Overview of proposed regulatory surface water MUSCRAT

Category	Description
Current model Version	MUSCRAT, Version 1.0
Theoretical basis	Regional summaries of multiple PRZM3 / EXAMS runs (25 per multi-state region)
Key assumptions	Same as for PRZM3 and EXAMS; regional soils, agronomics and weather selected based upon soils appropriate for crop
Key inputs and outputs	Compartment properties (physical and chemical), chemical input loadings (e.g. drift, run-off and erosion); chemical concentrations in compartments over space and time

Tier 3 modelling is currently performed using an automated shell around PRZM3 and EXAMS called MUSCRAT (Multiple Scenario Risk Assessment Tool) [6]. This model performs PRZM3 / EXAMS assessments using a range of soil and climatic inputs to provide a probabilistic evaluation of regional surface water impacts. Table 10 gives the key information for the MUSCRAT tool with respect to the models PRZM3 and EXAMS.

The USEPA are currently developing higher tier approaches to assess potential impacts of pesticides on surface water. Watershed modelling is being developed to support higher tier ecotoxicological risk assessments. Statistically-based water quality monitoring is also being developed to provide information on the concentrations in surface water and groundwater, which serve as drinking water.

In figure 5 an example is given of the deterministic output of PRZM3 / EXAMS related to different inputs as drift, run-off and erosion, while figure 6 shows the probabilistic output of the tool MUSCRAT.

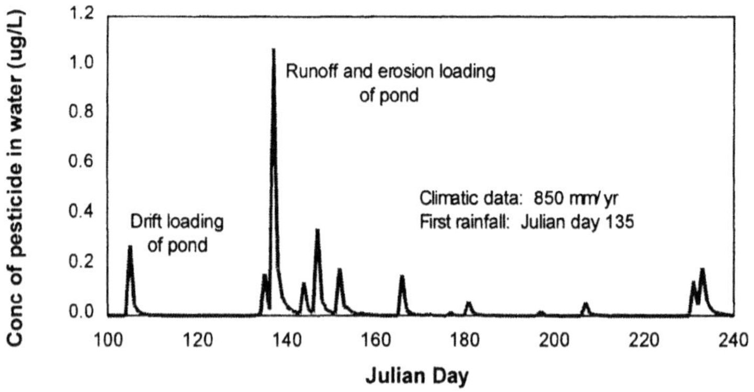

Figure 5. Example of Deterministic PRZM3 / EXAMS Output

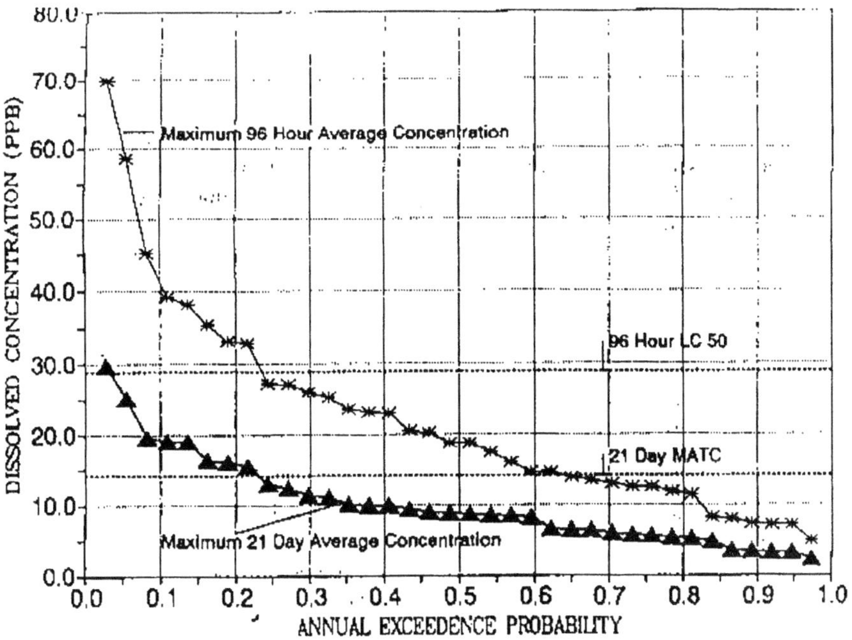

Figure 6. Example of Probabilistic MUSCRAT Output

3. Hazard Assessment

The evaluation of the intrinsic hazard that a chemical poses to various organisms can be evaluated either by various types of structure-based screening and database estimates or by directly measuring the hazard in the laboratory, simulated field or field studies. The USEPA currently requires a battery of aquatic and terrestrial hazard data to support registration of new pesticides. These data include studies of fish, aquatic invertebrates, marine species, aquatic plants, sediment-dwelling organisms, mammals, birds, bees and non-target plants [1]. Table 11 and 12 show an overview of the key aquatic and terrestrial hazard data required by USEPA.

Most hazard data for pesticides is generated from direct experimental studies of the chemicals with various test organisms. Models used in conjunction with these studies include dose/response models to determine various levels of effects, kinetic models to determine the degradation of pesticides on food sources for birds and a bioconcentration model for simulating uptake and depuration rates in fish.

TABLE 11. Aquatic hazard data required by USEPA

Species tested	Type of test and endpoint
Freshwater fish	Acute: 96 h LC50 (rainbow trout, bluegill sunfish)
	Chronic: partial or full life-cycle (fathead minnow)
Freshwater aquatic invertebrates	Acute 48 h EC50 (daphnia spp.)
	Chronic: 21 d life-cycle (Daphnia magna)
Marine / estuarine species	Acute: 96 h LC50 (sheepshead minnow)
	Acute: 96 h EC50 (crustacean, e.g. mysid shrimp)
	Acute: 96 h EC50 (eastern oyster)
Aquatic plants	Acute: 120 h EC50 (4 algal species)
	Acute: 14 d EC50 (1 aquatic macrophyte, e.g. lemna)
Sediment organisms	10 – 28 d LC50, NOEC (C. tentans)

TABLE 12. Terrestrial hazard data required by USEPA

Species tested	Type of test and endpoint
Mammals	Acute: LD50 (1 species, typically rat)
Terrestrial vertebrates (birds)	Acute: oral LD50 study (1 species)
	Acute: 5 d dietary study (2 species)
	Chronic: reproduction (1 species)
Bees	Acute: 48 h LD50, NOEL contact or residue study
Non-target plants	Vegetative vigour and seeding emergence EC50 (10 sp)

4. Risk Assessment

The current method of risk assessment used by the USEPA involves calculating ratios of exposure and hazard known as risk quotients [1]. For acute exposure, the exposure should be less than one-half of the defined hazard endpoint. For chronic exposure, the risk quotient should not exceed a value of 1.0. To provide a more detailed and informative level of risk assessment, the USEPA is currently developing risk assessment methodologies, which involve evaluating the probabilities of observing various levels of risk rather than using a single risk quotient. This approach, developed in part by a recent group called ECOFRAM (Ecological Committee on FIFRA Risk Assessment Methodologies) [7], involves generating probabilistic distributions of exposure, hazard and risk. Various types of regressions and mechanistic models have been recommended for the generation of the needed distributions. In addition to models, various types of refined experimental approaches have been recommended to permit more realistic assessment of the effects on chemical exposure to aquatic and terrestrial organisms. Figure 7 gives a generalised idea of the probabilistic approach when comparing probabilistic exposure calculations to probabilistic hazard information and how this combines to probabilistic risk. In figure 8 it is shown how the sensitivity of different organisms may vary as a function of concentrations.

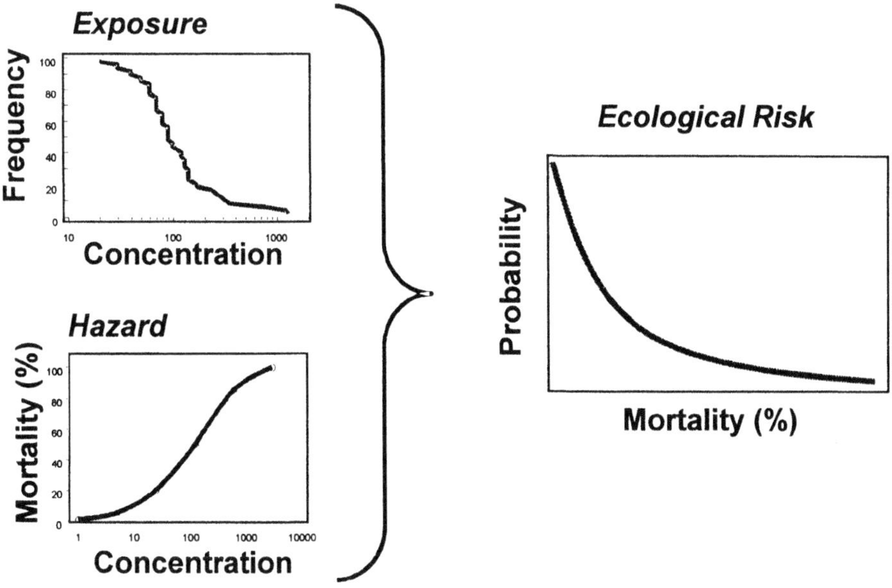

Figure 7. Example of Probabilistic Exposure, Hazard and Risk from ECOFRAM

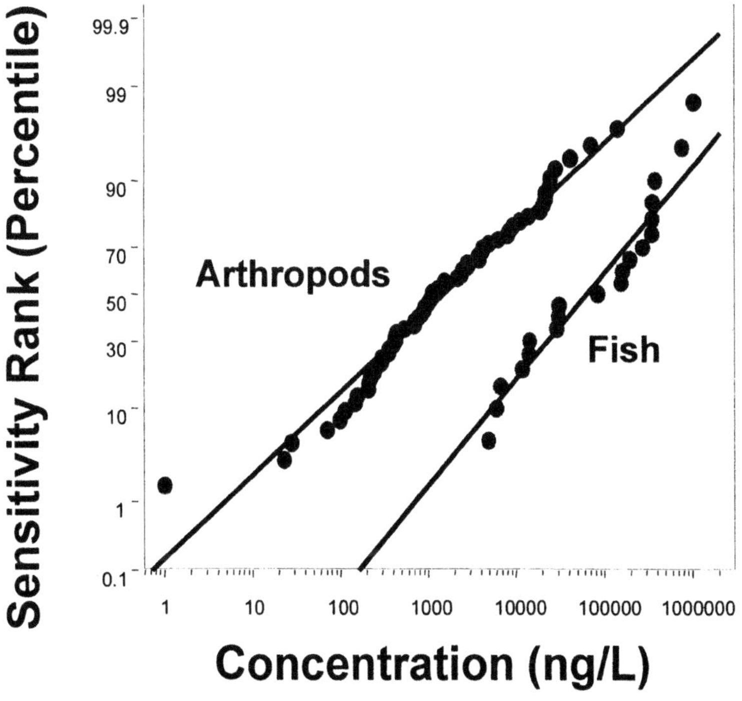

Figure 8 Example of Distribution of Species Sensitivities from ECOFRAM

5. Issues in Simulating Leaching, Runoff and Erosion

There are a number of current issues concerning modelling of impacts on groundwater via leaching of pesticides and surface water via runoff and erosion.

Leaching issues include development of appropriate algorithms to permit simulation of preferential flow, which is known to occur in a wide variety of agricultural soils. Another significant issue is the development of refined kinetics to permit the accurate representation of the degradation of parent and metabolites as a function of temperature, moisture, depth and time. Some newer models have included sorption kinetics to account for the increasing strength of sorption that normally occurs with time in soils.

A major issue in regulatory modelling of pesticides is the selection of appropriate chemical input values. Draft guidance on this topic has been generated by the USEPA and guidance should be completed by the end of 1999 [8]. When it is necessary to provide a more accurate simulation of the transient movement of water in the soil profile, models that use the Richards equation such as MACRO or PELMO should be used rather than PRZM3. However, these models are not routinely used for regulatory submissions in the USA to date. As the recommendations of ECOFRAM are implemented by the USEPA, probabilistic modelling will continue to increase in importance.

A key issue resulting from modelling of runoff, leaching and surface water is the selection of appropriate environmental fate data for use in modelling. Current regulatory studies are generally sufficient for most chemicals. However, for certain specialised situations, such as simulation of pesticide degradation in a rice paddy, it is occasionally necessary to perform additional studies to provide realistic data for use in appropriate models.

The simulation of drift is an important issue in all regulatory modelling. In the USA, the Spray Drift Task Force (SDTF) has developed a model called AgDrift [9] that is used to provide estimates of drift as a function of the type of spray equipment and worst-case atmospheric conditions. For higher tier modelling evaluations, it is appropriate to use more refined (i.e. more realistic) atmospheric conditions to determine the likely extent of drift from treated agricultural fields into non-target areas.

Current modelling approaches using GENEEC or PRZM3 and EXAMS assume that edge-of-field runoff and erosion directly enter adjacent surface water bodies. Natural buffer zones attenuate runoff and erosion by 30 to 50% or more and should be considered in higher tier evaluations [10].

Simulation of tile drainage is still an active area of research in many countries of the world. A recent review of experimental data indicates that the highest concentrations of pesticide in tile drainage are determined by preferential flow [11]. As a result, it is currently recommended that all modelling of tile drainage be calibrated to experimental data since no existing models are able to accurately predict preferential (or macropore) flow without calibration.

Ideally, surface water models should reflect the highly dynamic volumes and flow rates that result from storm events to provide realistic concentrations in surface water due to runoff, erosion, drift and drainage. However, for regulatory modelling, it is often sufficient to use streams and ponds with constant volume and flow rates to obtain initial estimates of the potential concentrations of pesticides in surface water.

Finally, it is appropriate that regulatory models should be validated against experimental data to ensure that the predictions generated by these models are acceptably accurate to support sound regulatory decision-making. Validation efforts have recently been performed in both the USA (FIFRA Exposure Model Validation Task Force, validating PRZM3 for runoff, erosion and leaching) [12] and a COST66 project in the EU, evaluating a wide range of models against a set of European field studies [13].

6. References

1. USEPA/EFED, (1998) "A Comparative Analysis of Ecological Risks from Pesticides and Their Uses: Background, Methodology & Case Study",. Current methods and proposals for comparative methods. Document provided on Model CD, November.1998

2. Barrett, Michael (1998). SCI-GROW (Screening Concentration in Groundwater). Manual and software provided on Model CD.

3. Carsels, R.F., J.C. Imhoff, P.R. Hummel, J.M. Cheplick, and A.S. Donigian, Jr. (1997). "PRZM-3, A Model for Predicting Pesticide and Nitrogen Fate in Crop Root and Unsaturated Soil Zones: Users Manual for Release 3.0", National Exposure Research Laboratory, U.S. Environmental Protection Agency, Athens, GA 30605. Manual and software provided on Model CD.

4. Parker, R.D., H.P. Nelson and R.D. Jones, (1995). "GENEEC: A Screening Model for Pesticide Environmental Exposure Assessment", in Water Quality Modelling, Proceedings of the International Symposium, ASAE, April 1995. Manual and software provided on Model CD.

5. Burns, Lawrence A., (1994). "Exposure Analysis Modelling System: Users Guide for EXAMS II Version 2.95", Environmental Research Laboratory, U.S. Environmental Protection Agency, Athens, GA 30605. Manual and software provided on Model CD.

6. Mangels, Gary, (1997). "Multiple Scenario Risk Assessment Tool (MUSCRAT), Version 1.0 (beta)". Available from American Cyanamid at mangelsg@pt.cyanamid.com

7. ECOFRAM (Ecological Committee on FIFRA Risk Assessment Methods), final report to be issued by end of 1999.

8. "Guidance for estimating metabolic degradation input parameters for GENEEC, PRZM, and EXAMS when estimating exposure in surface water", draft guidance issued by R. David Jones, USEPA/EFED, April 28, 1998. For information, contact jones.rdavid@epa.gov

9. AgDrift, a model developed by the Spray Drift Task Force (SDTF) in the USA. For information, contact Dr. David Johnson, Project Manager, Stewart Agricultural Research Services, Inc., Macon, MO USA (660) 762-4240.

10. Misra, A.K., J.L. Baker, S.K. Mickelson, H. Shang, (1996). "Contributing Area and Concentration Effects on Herbicide Removal by Vegetative Buffer Strips", *Transactions of the ASAE*, Vol 39(6):2105-2111.

11. Kladivko, L.C. Brown, and J.L. Baker, (1999). "Pesticide Transport to Subsurface Tile Drains in Humid Regions of North America", Report prepared for the American Crop Protection Association, June 22, 1999.

12. FEMVTF (FIFRA Exposure Model Validation Task Force), final report to be issued by end of 1999.

13. COST66 Model Validation Project, final report submitted for publication in *Pesticide Science*.

QUALITY ASSURANCE IN ENVIRONMENTAL MODELLING

GYULA DURA & ELIZABETH LASZLO
National Institute for Environmental Health of "Fodor József"
National Centre for Public Health
Gyali ut 2-6,
H-1097, Budapest, Hungary

Abstract

The different view of scientists and decision makers in risk assessment, the gap between needs and opportunities, the increasing number of models and their wide application by practical users with different background make necessary to ensure the quality and performance of modelling.
For this reason, it is very important to have a clear description of the use and the limitation of the model and the input data useability as well as the assumption and suggestions applied in the risk assessment process. To comply with the general quality requirements a risk assessor should manage the questions of data needs, variability, uncertainties inherent to each component of risk assessment. Statement of working hypothesis and applied models that improve understanding of numerical risk estimates is essential.

1. Introduction

Originally, environmental computer models have been developed and used as research tools. The potential usefulness of models for exposure and risk assessment is recognised by environmental and public health authorities and industry, which have to make decisions on regulation, hazard assessment and monitoring. While developing and using models, scientists describe environmental processes and usually improve our knowledge accordingly. Researchers know well the suggestions and limitations of the models and the uncertainties of the results. Decisions makers, however, use the same models mostly as management tools without deep understanding the environmental processes, adverse health and ecological effects and they may not be sufficiently aware of the limitations of the predictions.

J.B.H.J. Linders (ed.), Modelling of Environmental Chemical Exposure and Risk, 31–37.

The different view of scientists and decision makers in risk assessment, the gap between needs and opportunities, the increasing number of models and their wide application make necessary to ensure the quality and performance of modelling.

2. General Consideration of Modelling

In practice, there are two main areas of environmental modeling. The first is the assessment of the behavior and fate of chemicals at a given location, the purpose being to predict concentrations resulting from the local contamination. The second is the screening and regulatory control of new and existing chemical substances where the goal is to evaluate the potential exposure, typically on a much larger regional scale. These two areas of modeling require different input data.

The former requires relevant data and models applicable to the site. Mostly, local situations are complicated and information about the fate and behavior of chemicals in the local environment is limited. The second case requires definition of a generic environment which will be representative of the typical use or the life cycle of the chemical. This can be conducted at a screening or more advanced level. At a screening level a "standardized" environment is envisaged. Multi-media partition models are of particular interest at this level. At an advanced level the same simple, or more sophisticated multi-compartment models, can be used but with more comprehensive input data.

3. Selection of Models

Before selecting a model, the fundamental problem is to define exactly the question a model is intended to answer and the level of accuracy required.

Once this has been decided, it will be possible to select a model which includes all the relevant processes and environmental situations. In the process of selection of appropriate models it is necessary to find answers to the following frequently asked questions:

- What is the nature of the problem and the time and spatial scale for risk assessment?
- What are the management goals with risk characterisation?
- What are the populations of concern and receptor characteristics?
- How can environmental parameters be taken into account?
- Is it correct to take a so-called "worst-case approach"?
- How far can monitoring data be taken into account in the application of the models?

Once having chosen a suitable model and obtained data it is still easy to misinterpret the predicted results if the underlying principles are not properly understood. Furthermore there are some concerns about the uncertainty inherent in risk assessment and preference of measurement on the contrary of modeling. For this reason, it is essential to have a transparent description of the use of the model and the input data required in compliance with quality assurance requirements. To comply with the general quality requirements a risk assessor should manage the questions of data needs, uncertainty, variability and the risk assessment software itself.

4. Data needs

After selection of the model one of the most important moments in conducting risk assessments is to explicitly identify each default option when used. It suggests the clear and consistent descriptive parameters related to a given environmental situation. It is very important to give much more significance if you change default options. In this way we can provide enhanced guidance to the public and reduce the possibility of actions on a sudden impulse and undocumented shift that could decrease credibility of obtained results,

There is a general lack of exposure and toxicity data needed to assess the chemical risks. Therefore it is important to define the types, quantities, and quality of data needed. It is recommended to compile an inventory of the physicochemical, toxicological, ecotoxicological, epidemiological and regulatory toxicological literature data on each chemicals in question. After identifying data gaps data might be generated by other methods including QSAR techniques [1]. When assumptions are made, the source and general considerations used to develop the assumptions according to analogy, professional judgement should be described.

Considerable attention should be turned to mobile and diffuse sources of hazardous chemicals, as they may be more important in some cases than the point sources of pollution. The concentrations of contaminants to which a population may be exposed represent the most important determinant of risk. The numerical estimate of chemical concentration in an environmental sample is often accompanied by "data qualifier" that provides guidance to interpreting numerical values for quantification limit and "non-detects" [2]. For monitoring data, there should be a control of the data quality objectives as they are relevant to risk assessment.

In exposure scenarios explicitly all direct and indirect routes of exposure to pollutants should be reflected, including ingestion, inhalation and dermal absorption. Appropriate selection of human physiological parameters with dosimetric significance is a precondition of the correct assessment.

Because pollutants emitted into the environment may be transferred in other exposure media, appropriate statistical procedures are used for aggregating risks from exposure to multiple compounds and from exposure to chemicals via multiple media and routes.

5. Variability

Although the health risks from a given pollutant may vary appreciably among individuals and among populations, such variability has generally received little consideration in risk assessments, in part because of limitations in the availability of relevant data. To improve this situation we have to refine estimates of risks to individuals and population groups. Variability comes from true heterogeneity in characteristics such as dose-response differences within a population, or differences in contaminant levels in the environment. The values of some parameters used in an assessment change with time and space, or across the exposed population. Assessments should address the resulting variability in doses received by members of the target population. Individual exposure, dose, and risk can vary widely in a large population. The main tendency can characterise the variability in exposure, lifestyles, and other factors that lead to a distribution of risk across a population. It is necessary to assess risks to infants and children specifically whenever their risks appear likely to exceed those of adults and distinguish rigorously between individual variability and other sources of uncertainty in each component of risk assessment.

6. Uncertainty

Uncertainty represents lack of knowledge about factors such as contaminant levels or adverse effects, which may be reduced with additional studies. Generally, risk assessment bears several categories of uncertainty, and each deserves consideration.

Measurement of uncertainty reflects the common error that accompanies scientific measurements and standard statistical techniques can be used to express the magnitude of this uncertainty.

Sampling uncertainty is also inherent in environmental exposure and risk assessment, and it should be addressed.

There are similar uncertainties associated with the use of scientific models, e.g., dose-response models, models of environmental fate, transfer and transport. Evaluation of model uncertainty would consider the state-of-the-art for the model and the scientific basis for the model validation.

Other kinds of uncertainty rise from data gaps that is, estimates or assumptions used in the assessment. Often, the data gap is wide, such as the absence of information on the effects of exposure to a chemical on humans or on the biological mechanism of action of an agent.

The risk assessor should include a statement of confidence that reflects the degree to which the risk assessor believes that the estimates or assumptions adequately fill the data gap. For some common and important data gaps, specific risk assessment guidance provides basic assumptions or generic values [3]. Risk assessors should carefully consider all available data before relying on assumptions.

Unless each of the various sources of uncertainty typically inherent in a risk assessment is adequately identified and explained, a decision-maker acting on the assessment may not know the extent of conservatism, if any, provided by the assessment. Therefore it is advisable to perform formal uncertainty analyses, which could show where additional data might resolve major uncertainties. We have to emphasise the limits of relevant scientific knowledge, as well as the need to identify and minimise errors of underestimation or overestimation. We recommend to quantify the uncertainties inherent to each step of the risk assessment process and inform risk managers about the uncertainties inherent to risk assessments, and not only give them a single point estimate or range of numbers for characterising a risk. The facilitation of an uncertainty analysis represents a possibility to ensure quality. Computer models should support the uncertainty analysis or cooperate with special programmes such as Monte Carlo analysis.

7. Qualitative Risk Description

Sometimes risk assessors simplify the discussion of risk characterisation by speaking only of the numerical components of an assessment. A simplified numerical presentation of risk, referring to the numeric risk quotients and weight-of-evidence classification of cancer risk, is always incomplete and often misleading. For this reason, the NRC [4] and EPA risk assessment guidelines [5] call for "characterising" risk to include qualitative information, a related numerical risk estimate and a discussion of uncertainties, limitations, and assumptions.

Qualitative information on working hypothesis and applied methods is an essential component of risk characterisation. For example, specifying that animal studies were used in an assessment underlines that the risk estimate is based on assumptions about human response to a particular chemical rather than human data. Qualitative descriptions of this kind provide elementary information that improve understanding of numerical risk estimates. These uncertainties are expected in scientific studies. Such uncertainties do not reduce the validity of the assessment.

In many cases, risk assessors must choose among available data or assumptions in evaluating risks. Examining the impact of selected, plausible data on the quantified risk and conclusions of the assessment is the crucial component of the uncertainty discussion. Risk assessor, using best professional judgement, should outline the strengths and weaknesses of the plausible alternative approaches. The degree to which uncertainty is addressed and explained depends largely on the purpose of the assessment and the data, information and resources available. Nobody expects an assessment to evaluate all possible exposure scenarios for every potential contaminant, to examine all vulnerable population groups at risk. Rather, the qualitative risk description should reflect the complexity of the risk assessment, with the level of effort for analysis and discussion of uncertainty corresponding to the level of the assessment.

8. Quality Requirements for Risk Assessment Software

In order to be confident in the applicability of the models, an evaluation of the model structure and of the software is required. Quality requirements for software products

should be specified in detail [6].
Quality assurance deals with the organisational structure and the conditions, under which laboratory or field studies are planned, completed, and evaluated, as well as the recording and reporting of the investigation. A similar approach is desirable for the generation and use of computer programmes for risk assessment, for which Good Modelling Practice (GMoP) should also be established. The bases for this are quality criteria for models for exposure and risk assessment, which can only be found in few publications [7, 8, and 9].
The description of the software for risk assessment is not as important as for standard office software. However, it should still be available in order to look into the technical background and areas (indication) of application and its limitation or possible communication with databases.
The well-documented manual should contain both functional description (installation, operation, etc.) and professional references (theoretical description of the models). All of the equations in the documentation should be presented with references of the original literature. Ranges of parameters and degree of correlation should also be expressed. It is evident that the programme must calculate correctly. All programme settings and especially the input parameters should not be unnecessarily limited. This gives the programme a wide range of applications and makes it operable for general and specific environmental conditions. Results of programme running should be concise. Choice of presentations in forms of tables, graphs, or text, key-data, background informations are an asset.
It must be clear to users at all times which calculations are being carried out and how individual models can be linked together. This transparency is achieved by free insight into equations, variables, units and the logical modules of the models. The transparency of the models is an elementary requirement for the acceptance of the software.
The programme should not be more sophisticated than necessary. If the number of equations and parameters used and other conditions are kept to a minimum, the whole programme is easier to understand, thus contributing considerably to its transparency.
Because of the purpose of risk assessment software as a decision support tool for experts, a certain amount of scientific knowledge is required to use these programmes. But even experts may not know all ranges of each parameter. For this reason it is also an advantage for a programme to support users when entering data and applying the models. Important operational requirements within the framework of quality assurance are:

- warning messages: if nonsense data is entered or if the regression range is out of the regression model, the programme should deliver the appropriate message,
- interdependencies between parameters should be monitored,
- conversion of units and
- export and import of raw and estimated data should be provided.

9. Concluding Remarks

The scientific approach for assessing chemical risks to environment and human health is fundamentally sound, but at the same time there is a need to amend nearly every step of the risk assessment process. Quality assurance could improve the reliability of the risk assessment process.

Risk assessment has been characterised on one hand as holistic-science and on the other hand science providing solutions to local, site-specific problems. It is certainly controversial and complicated, requiring scientists to understand mathematical models and mathematicians to understand different areas of environmental sciences. In essence, risk assessment requires that we help decision-makers in the application of findings on the adverse effects of chemicals.

As the models become more complicated and the practical users come from different backgrounds other than the traditional modelling, it is likely that underlying assumptions regarding biological effects or the behaviour of chemicals in environment will not be taken into consideration completely. This is of great concern because slight changes in the assumptions might cause high influence on the predictions obtained by the model.

The key idea in using models is to describe and understand the environmental processes that go into the model. The level of knowledge in modelling and good modelling practice will greatly determine the quality of the prediction.

10. References

1. Structure – activity relationships in toxicology and ecotoxicology (1986) ECETOC, Monographs No.8.

2. Risk Assistant for Windows 3.1 User's guide. Hampshire Research Institute, Inc. 1995.

3. EU TGD (1996) "Technical Guidance Document in Support of Commmission Directive 93/67/EEC on Risk Assessment for New Notifield Substances and Commission Regulation (EC) No 1488/94 on Risk Assessment for Existing Substances", European Commission

4. National Research Council (1983) Risk Assessment in the Federal Government: Management the Process.

5. U.S. EPA (1992) Guidelines for Exposure Assessment **57** Federal Register, 22888 – 22938, May 29

6. DIN ISO/IEC (1995) 12119: 1995-08, Berlin

7. Estimating environmental concentrations of chemicals using fate and exposure models (1992) ECETOC, Technical report No.50.

8. Schwartz S., Berding V., Trapp S., Matthies M. (1998) Quality criteria for environmental risk assessment software. Using the example of EUSES. *ESPR –Environ.Sci.& Pollut. Res.* **5**, 4. 217 – 222.

9. Portier C. J. , Lyles C. M.(1996) Practicing Safe Modeling: GLP For Biologically Based Mechanistic Models. *Environ. Health Perspect.* **104,** 8. 806

POSSIBLE APPROACHES FOR PESTICIDES ENVIRONMENTAL IMPACT MANAGEMENT

M.G. PRODANCHUK & ALEXANDR P. KRAVCHUK
Medved's Institute of Ecohygiene and Toxicology
6 Heroiv Oborony Street,
Kiev
Ukraine

1. Introduction

Pesticides are new artificial environmental factors that may cause dangerous influence on organism. Due to their highly biological activity, and permanently direct and/or indirect contact with widespread human population these factors require an especially complex methodological approach to their hazard assessment.

Evolution of opinions on the requirements to such an approach went through the following stages:

- 1st stage: Determination of an individual functional threshold. In this stage there is a need for the prevention of an acute poisoning and for the determination of acute toxicity for this purpose.
- 2nd stage: Ontogenetic threshold. Understanding of the necessity for the evaluation of the general biologic impact on organisms (investigation of cumulating, subchronic and chronic toxicity, allergenic properties, delayed neurotoxicity, carcinogenic activity) and on the next generation (mutagenic activity, reproductive toxicity) of organisms is the main aim of this stage.
- 3rd stage: Population threshold. There is a necessity for the evaluation of the impact on populations (risk assessment, combine action of other environmental factors, etc.).
- 4th stage: Phylogenetic threshold. In this stage there is a necessity for the assessment of potential general changes in the biosphere (elimination of the most sensitive persons as a consequence of mutagenic action and the selective impact of global pollutants (i.e. "environment hyperestrogenisation phenomenon", etc.)

Thus, the hazards of pesticide impact can be divided into two conditional categories: hazard of acute toxic action at direct contact and hazard of indirect chronic action through the environment (figure 1).

J.B.H.J. Linders (ed.), Modelling of Environmental Chemical Exposure and Risk, 39–46.

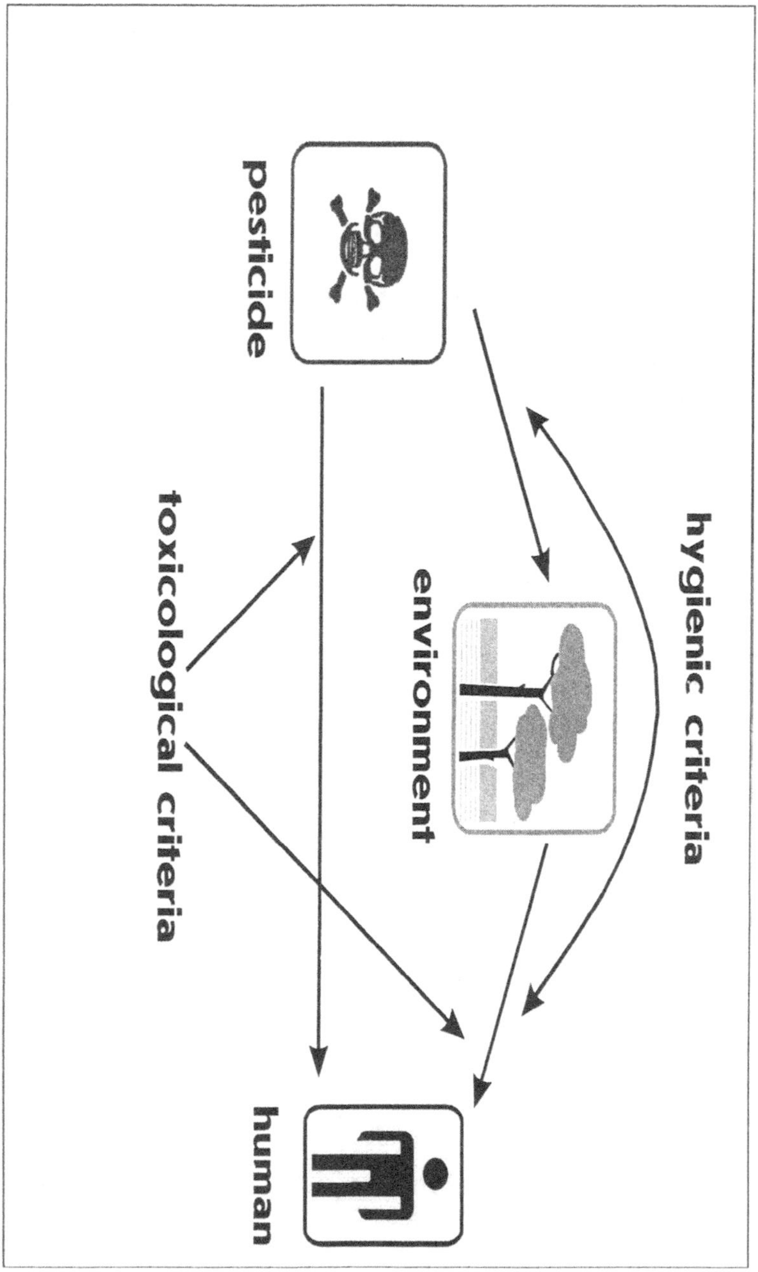

Figure 1. *The direct and indirect impact of pesticides on man*

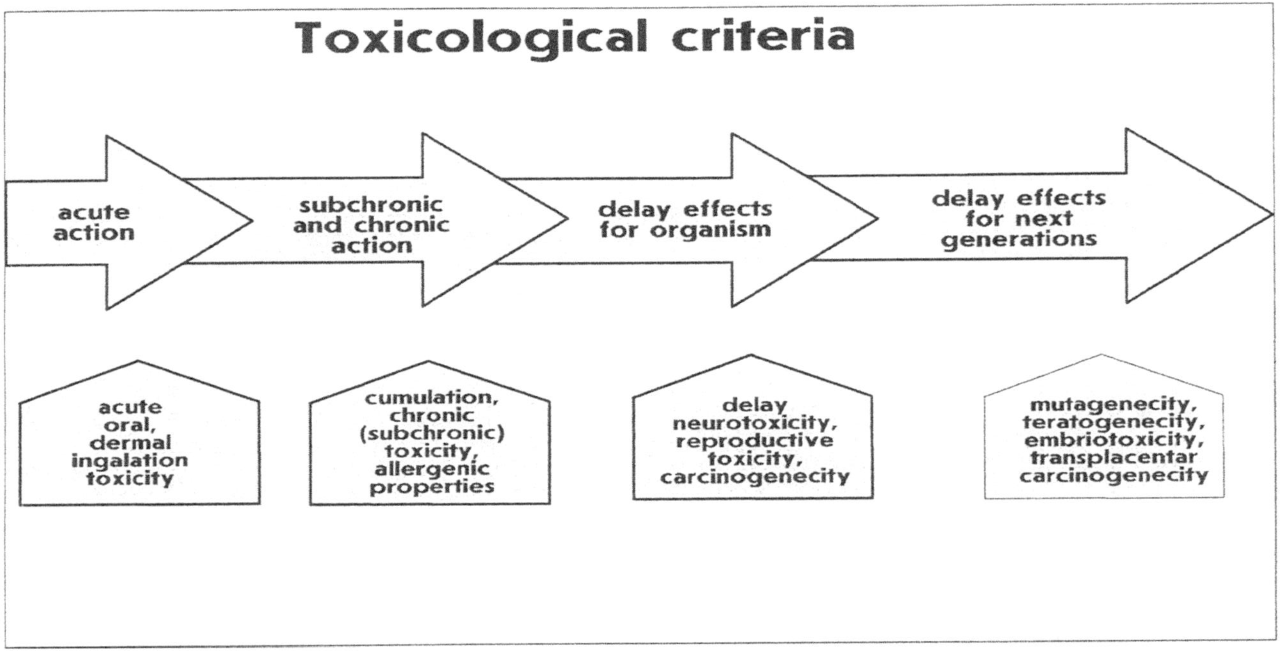

Figure 2. Main toxicological criteria in relation to the effects in a timely perspective

2. Criteria Approach

The main toxicological criteria are reflected on figure 2.

According to the generally accepted approaches, the most important integral toxicological criterion is the acceptable daily intake (ADI) which is calculated based on the lowest no effect level of the limited criteria. The ADI is the maximum dose of a pesticide, averaged as a daily intake that could not cause illnesses, or health deviations, or have effects on future generations during the whole lifetime.

It is obvious that different pollutants have a different priority for each region and can have different effects in each case. This is the reason for using the hygienic criteria (figure 3) in the hazard evaluation of pesticides.

3. Exposure Assessment

The aforementioned criteria reflected potential hazard. And we need to calculate the actual dose of pesticides that organism receive through water, air and food for the evaluation of their real hazard.

We are considering the possibility to attain its ADI as the basic criterion of real hazard of each pesticide.

The general formula to account for the actual pesticide dose (Da) is:

$$Da = \frac{D_1 + D_2 + D_3}{I} \qquad (1)$$

Where:

D_1	=	Daily pesticide dose, received through food, mg;
D_2	=	Daily pesticide dose, received through drinking water, mg;
D_3	=	Daily pesticide dose, received through ambient air, mg;
Ì	=	Average human weight (in the given group of the population), kg.

The common pesticide dose, received through food (D_1), is calculated by summation of the amount which has been found in separate products that reflected an average diet for the consumption of the selected population:

$$D_1 = \sum \tilde{N} \cdot Q_i \cdot (1 - q_i) \qquad (2)$$

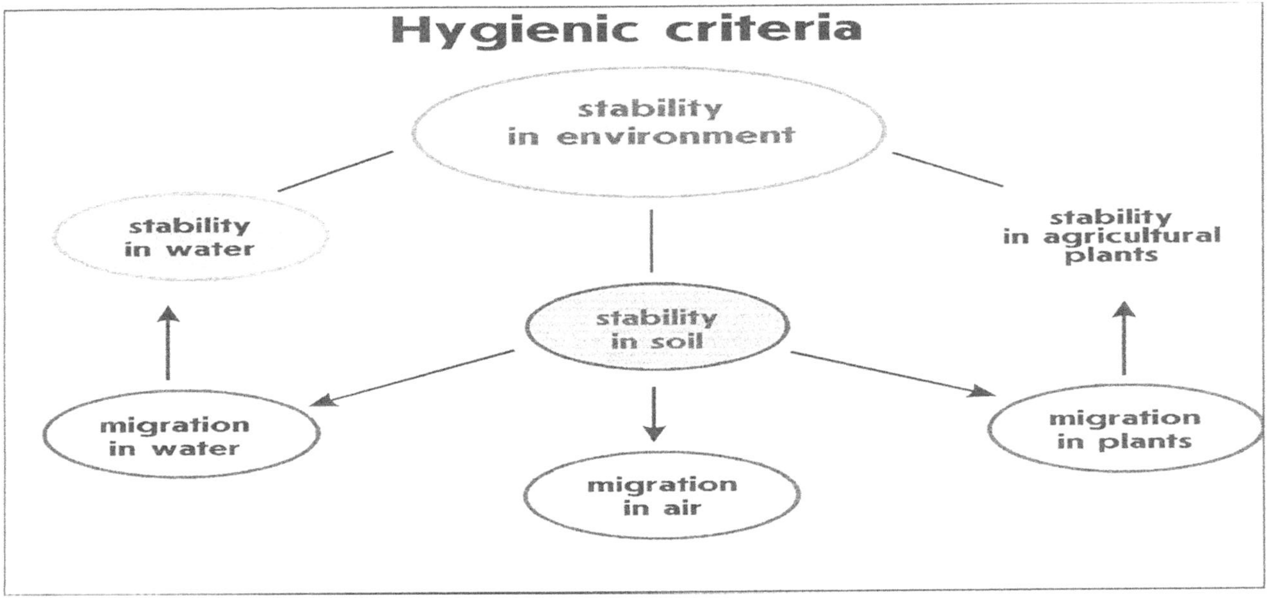

Figure 3. Hygienic criteria in pesticides'hazard evaluation

Where:

$\tilde{N}_i$	=	Concentration of a preparation in product, mg/kg;
Q_i	=	Average amount of product in daily diet consumption, kg,
q_i	=	Factor of pesticide destruction during the cooking process of the product depending on the available information; if not available q = 0.

The intake of pesticide with drinking water (D_2) can be described as follows:

$$D_2 = \tilde{N}w * R \qquad (3)$$

Where:

$\tilde{N}_w$	=	Concentration of a preparation in drinking water, mg/L;
R	=	Daily intake of water, L.

To define the dose of pesticide received by the human organism through ambient air (D_3), the following formula is used

$$D_3 = T_o * \tilde{N}_v * V * b_v \qquad (4)$$

Where:

T_o	=	Oral/inhalation factor for given pesticide;
$\tilde{N}_v$	=	Pesticide concentration in ambient air, mg/m^3;
V	=	Daily human respiration value, m^3;
b_v	=	Factor of pesticide absorption à in respiratory tract.

Analysis of data obtained in different agricultural regions shows, that D_1 forms 70-90 % of Da, D_2 about 10-30 % and D_3 circa 3 - 10%.

The possibility of pesticides to attain the ADI (actual exposure index) is estimated through the comparison of Da with the ADI:

$$AEI = \frac{Da}{ADI} \qquad (5)$$

Where:

AEI	=	Actual exposure index.

4. Examples

The level of AEI may be taken as a basis for the determination of pesticide pollution priority. To illustrate this we have used three important pesticidal pollutants in Ukraine. It concerns the persistent organic pollutants: DDT and HCCH and the herbicide with greatest scale of application: 2,4-D (tables 1-3).

All figures ware collected in a typical agricultural area in the north part of Ukraine and reliably reflect a real situation in the above region. For calculation of actual pesticide dose (Da) two food products with most homogeneous levels of pollution (milk and grain cultures) were chosen.

TABLE 1. AEI calculation for 2,4-D

	The contents of pesticide	Daily consumption	Daily dose (mg/kg)
Water	0.0001 mg/l	3 l	0.0003
Air	0.0001 mg/m^3	20 m^3	0.002
Milk and dairy products (in recalculation on milk)	0.0005 mg/kg	1.225 kg	0.000613
Grain products (in recalculation on flour)	0.009 mg/kg	0.38 kg	0.00342
D_a	0.000106		
ADI	0.0001		
AEI	1.06		

TABLE 2. AEI calculation for DDT

	The contents of pesticide	Daily consumption	Daily dose (mg/kg)
Water	0.0001 mg/l	3 l	0.0003
Air	0.003 mg/m^3	20 m^3	0.06
Milk and dairy products (in recalculation on milk)	0.004 mg/kg	1.225 kg	0.0049
Grain products (in recalculation on flour)	0.005 mg/kg	0.38 kg	0.0019
D_a	0.001118		
ADI	0.02		
AEI	0.06		

TABLE 3. AEI calculation for HCCH

	The contents of pesticide	Daily consumption	Daily dose (mg/kg)
Water	0.0001 mg/l	3 l	0.0003
Air	0.004 mg/m^3	20 m^3	0.08
Milk and dairy products (in recalculation on milk)	0.002 mg/kg	1.225 kg	0.00245
Grain products (in recalculation on flour)	0.002 mg/kg	0.38 kg	0.00076
D_r	0.001392		
ADI	0.001		
PEI	1.39		

The final table of AEIs looks as follows (see table 4.).

TABLE 4. AEI overview

Pesticide	AEI
DDT	0.06
HCCH	1.39
2,4-D	1.06

5. Conclusion

As described in table 4, the possibility to attain ADI of DDT is negligible. Therefore, the monitoring priority for DDT is low. At the same time, there is a real opportunity to the attain ADI of HCCH . The problem of environmental pollution by this compound is obvious. Such a situation can be explained by the wide application HCCH in former USSR (its application was banned in 1987 but it was still in use till middle 90-th). DDT has not been in use in Ukraine since 1969.

And, at last, environmental pollution of 2,4-D seems also realistic. This fact comes as no surprise if the real quantities of its application are considered. In 1999 its projected use in Ukraine agriculture was 7450 thousand tons on 5530 thousand ha (total agricultural area is 28890 thousand ha).

It is possible to divide the pesticidal pollutants into two conditional categories based on AEI approach, which is indicated in table 5.

TABLE 5. Criteria for categorisation

if AEI more or equal 1	undoubtful hazardous pesticides
if AEI less that 1	low hazardous pesticides or pesticides which need to use additional criteria for their hazard evaluation

It is obvious that urgent actions should be undertaken in order to reduce the potential impact of pesticides included in first category.

PESTICIDE LEACHING MODELLING VALIDATION

A Recent European Experience

MARNIK VANCLOOSTER
Department of environmental sciences and land use planning
Unité Génie Rural
Université catholique de Louvain
Place Croix du Sud 2, BP2
B-1348 Louvain-la-Neuve
Belgium

Abstract

Testing of pesticide leaching models is important in view of their increasing use in pesticide registration procedures in the European Union. This paper summarises the major conclusions and recommendation of a recent experience on pesticide-leaching models and modelling validation. Twelve models simulating the vertical 1-dimensional movement of water, solute, heat, and, in particular, pesticides, through the soil profile were used by 36 different model users. The models differ in terms of modelling concepts. Four quality data sets were identified to perform the analysis. The data set included field and lysimeter data, collected in the Netherlands, Germany, Italy and the United Kingdom. Non-structured as well as structured soils were available in the data set. To elucidate the ability to model correctly water transport, solute transport, heat transport and pesticide transport in soils, a stepwise evaluation approach was followed. Splitting up the experimental data set enabled us to quantify the calibration capability and the prediction capability of the models. The simulations were performed by different model users enabling us also to characterise output variability in terms of user dependent interpretation of the model input and parameters. This paper resumes the major conclusions and recommendations. Details of the modelling exercise are given in [30].

1. Introduction

Leaching of chemicals from the root zone of agricultural crops continue to endanger the long-term groundwater quality in agricultural areas all over the world. The leaching is determined by a range of biological and physico-chemical processes occurring within the soil-root system, and by the land use and climatic settings at the soil-atmosphere boundary. All these variables vary tremendously in time and space, which makes the

J.B.H.J. Linders (ed.), Modelling of Environmental Chemical Exposure and Risk, 47–59.

characterisation of the leaching behaviour in time and space a challenging, but complicated, task.

In comparison to experimental field studies, mathematical modelling allows easily to explore the theoretical leaching behaviour for a wide range of different scenarios. In the past, leaching modelling was mainly done within an academic framework, envisaging to explore the sensitivity of different soil, climate and land use parameters on leaching behaviour. Nowadays however, leaching modelling has also been introduced in decision-making, and is used by regulatory officers to assess risks associated with the use of crop protection products in agriculture. As part of the recent efforts to protect water resources in the European union e.g., harmonised registration procedures for crop protection products are developed by the different member states. These procedures are based on uniform principles and places great importance on the use of 'validated' mathematical models to calculate environmental concentrations of product residues for different realistic scenarios.

Given the wide range of possible conditions, which can be met in practice, there is a wide range of validity needed to accept leaching modelling as a versatile and credible risk assessment tool. Compared to the number of products and the number of scenarios, for which risks need to be accessed, only limited comprehensive studies on the validity of the pesticide leaching modelling are presented in the literature. In order to improve this validation status, an attempt was made to carry out such a study in Europe [30]. Researchers and model users, model developers and data set providers, were invited to co-ordinate their efforts and make a statement on the actual validity of pesticide leaching modelling. To this end, 4 high quality data sets on pesticide leaching were made available and compared to results obtained from 12 different leaching models as used by 36 model users. The implemented procedure allows not only to characterise differences in performances between different models, but also between different model users, and this for a range of, only, but yet 4, realistic European scenarios. This paper gives a short synthesis of the major results of this study. Details can be found in Vanclooster *et al.* [30]. The validation of the modelling was carried out to help the scientific community to identify the strong points, but also the bottlenecks and shortcomings, in the fast evolving modelling technology. Yet it should at the outset of this paper already be emphasised that the conclusions and recommendations are only valid for the time being. We may expect that the conclusions and recommendations presented below should be revised and updated as new modelling approaches and modelling practices will be presented in future.

2. Materials and methods

2.1 PESTICIDE LEACHING MODELLING AND MODELLING VALIDATION PROCEDURE

Mathematical modelling is the process of representing a real system, delineated in time and space, with a mathematical analogue, driven by model parameters and model input. Available models can be classified in terms of the spatial scale at which the model is operational (global scale model, catchment scale model, field scale model, point scale model, pore scale model), the temporal scale at which the model operates (instantaneous model, daily model, yearly model, long term model), the target user (engineering model, scientific model), the numerical technique that is considered (analytical model versus numerical model), and the way how input and parameters are specified (deterministic model, stochastic model).

Any model use is subjected to error due to biased conceptualisation within the model structure, wrong model parameterisation and input estimation. It is therefore highly recommended that a mathematical model and the modelling (i.e. the model and the way the model is used by a specific user) are subjected to a detailed analysis of its validity. Within our study, we adopted validation definitions as considered by the FOCUS working group on Leaching Models [5]. The FOCUS working group (FOrum for the Co-ordination of pesticide fate models and their Use) envisages to co-ordinate pesticide leaching modelling in Europe, with particular attention to the use of leaching models in the context of pesticide registration. The definitions proposed by this working group consider a model validated if the model has gone successfully through a validation process for a specified range of validity. A validation process is the action of comparing model output with data independently derived from experiments or observations of the environment. The range of validity corresponds to the part of the reality to which the validation of a model applies. The validation status of the model is the extent to which a model has successfully been validated within its range of validity. Note that the definitions above do not specify any criterion to accept our reject a modelling exercise. In addition, the definitions implicitly consider the model and the way the model is used by a specific user, i.e. modelling validation.

A detailed description of available models to simulate pesticide leaching has been reviewed in depth by Jury and Ghodrati, [16], Jarvis *et al.* [15], Thorsen *et al.*, [23]; and Vanclooster *et al.* [30]. The interested reader is also referred to the compiled data bases of leaching models such as reported by Camase [7], FOCUS [5] or COST66 [30]. The models that have been used in our exercise are given in Table 1. For a more detailed description of the models, the reader is referred to the references listed in the table. All models of Table 1 calculate pesticide leaching as a 1-D flow process within the unsaturated soil root zone of the crop. The selected models ignore the horizontal flow component that may occur in the unsaturated zone. Neither, do they appeal to any stochastic theory to consider a full 3-D description of the flow phenomenon. The models operate at least on a daily time step and integrate the daily leaching rate to obtain the total leaching amount. We further observe that both mechanistic leaching models (PESTLA, PESTRAS, MACRO, LEACHP, WAVE, SIMULAT, CRACKP) and functional, yet more empirical, models (PRZM2, VARLEACH, PLM, PELMO, GLEAMS) are considered. The former type of models is traditionally used in an academic context, whereas the latter type of models is more popular for registration and management applications. In addition, models are retained that explicitly account for preferential flow through macropores (MACRO, CRACKP, PLM).

TABLE 1. Models and major characteristics considered within the modelling test

Subject	CRACKP	GLEAMS	LEACHP	MACRO	PELMO	PESTLA	PESTRAS	PLM	PRZM2	SIMULAT	VAR-LEACH	WAVE
Water flow, governing equation	Capacity type	Capacity type	Richards equation	Richards equation	Capacity type	Richards equation	Richards equation	Capacity type	Capacity type	Richards equation	Capacity type	Richards equation
Solute flow governing equation	Convec-tive	Convec-tive	Convec-tion Dispersion Equation	Convec-tion Dispersion Equation	Convec-tion	Convec-tion Dispersion Equation	Convec-tion Dispersion Equation	Convec-tion	Convec-tion	Convection Dispersion Equation	Convec-tion	Convection Dispersion Equation
Preferential flow	yes	no	(yes)	yes	no	no	no	(yes)	no	yes	no	(yes)
Sorption	Linear	Linear	Freund-lich	Linear	Freund-lich	Freund-lich	Freund-lich	Linear	Linear	Freund-lich	Linear	Linear
Degrada-tion	1st order	1st order	1st order	1st order	1st order	1st order	1st order	1st order	1st order	1st order	1st order	1st order
Plant uptake	no	yes	yes	yes	yes	yes	yes	yes	yes	no	yes	yes
Volatiliza-tion	no	no	yes	no	no	no	yes	no	no	no	no	no
Reference	[3]	[21]	[8]	[15]	[17]	[4]	[11]	[20]	[25]	[1]	[14]	[28]

In order to implement a validation method, conform to the definitions proposed above, a protocol was implemented consisting of two subsequent phases. In a first phase, the model's fitting capacity is analysed. In the second phase, the ability of the model to extrapolate or to predict, is analysed. Calibration in this context is considered as a parameter estimation technique. During calibration, an object function is optimised such that the difference between model calculated and observed pesticide concentration is minimised. Such an optimisation can be done by changing the model input parameters by trial and error or by more elaborate statistical optimisation techniques. During the second phase, no re-adjustment of the model parameters is done. At this stage, independent data are used to test the model performance. It should be noted that, in a regulatory context, calibration as suggested here will in general not be possible, because it requires several field experiments for the same pesticide/soil combination that are usually not available. In pesticide registration procedures, a first assessment is made on the basis of pesticide properties measured under well-defined laboratory conditions. So for decision makers, it is crucial to know the potential of a model to predict pesticide behaviour without calibration. Another complication is that pesticide models consist of sub-models for simulating water flow, solute flow, heat flow, and pesticide behaviour. A method of testing should, therefore, be pursued allowing us to evaluate all sub-models in subsequent steps. Given these dilemma, the following pragmatic solution was applied in our model testing exercise. First, test the whole model for uncalibrated parameter values and report the results. Then, calibrate the sub-models in the sequence water flow, solute flow, heat flow, pesticide behaviour; and finally test the whole model with calibrated parameter values if data are available.

In model testing, particular attention needs to be paid to the target quantity to be considered. The sub-models for simulating water flow, solute flow, heat flow, and pesticide behaviour could be tested using system-state variables such as the soil moisture content, the soil solute concentration, the soil temperature, and the soil pesticide concentration. Alternatively, balance terms such as the cumulative soil water drainage or the cumulative amount of pesticide leached can be used. The balance terms may be more appropriately analysed in a risk analysis context. Yet the balance terms are hard to measure on a field scale, which makes the test of pesticide leaching models at this scale cumbersome. Correct cumulative amounts leached can be inferred from the lysimeter data sets. However, flow behaviour in lysimeters does not reflect the complete variability of pesticide flow that might occur in the field, and bias may result in the risk assessment. To cope with aforementioned dilemma, a pragmatic solution was proposed. For the field data sets, the simulated state variables are compared with the field measured state variables in space and time. Adopting such an approach, one tacitly assumes that if the model matches measured state variables in space and time, the model will have a large probability of simulating the balance terms appropriately. For lysimeter data sets, the different models calculated balance terms that were compared to the balance terms measured in the lysimeters. In this local scale approach, field heterogeneity - a phenomenon that undoubtedly dominates pesticide fate and transport - is not explicitly accounted for during the test.

The appropriateness of the model calculation can be either assessed by modelling statistics or simulation graphics. When both system output and model output are considered to be deterministic, the arithmetic difference between the measured and calculated output is the most straightforward measure of the model deviation. The large amount of data however enhances the use of more comprehensive model deviation indicators. A selection of statistical criteria, used in the current modelling exercise, is given in table 2.

TABLE 2. Selection of statistical model indicators

Overall fit		
Maximum error	$ME = \max\lvert P_i - O_i \rvert, \quad \forall i$	Loague *et al.*, 1988
Root mean square error	$RMSE = \frac{100}{\overline{O}}\sqrt{\frac{\sum (P_i - O_i)^2}{n}}$	Loague *et al.*, 1988
Modelling efficiency	$EF = \frac{\sum (O_i - \overline{O})^2 - \sum (P_i - O_i)^2}{\sum (O_i - \overline{O})^2}$	Loague *et al.*, 1988
Nash-Sutcliffe coefficient	$CNS = 1 - \frac{\sum (O_i - P_i)^2}{\sum (O_i - \overline{O})^2}$	Nash and Sutcliffe, 1970
Average difference	$AVDIF = \sum \frac{(O_i - P_i)}{n}$	Nash and Sutcliffe, 1970
Total Soil Residue		
Coefficient of residual mass	$CRM = \frac{\sum O_i - \sum P_i}{\sum O_i}$	Loague *et al.*, 1988
Distribution of model error		
Coefficient of determination	$CD = \frac{\sum (O_i - \overline{O})^2}{\sum (P_i - \overline{O})^2}$	Loague *et al.*, 1988

2.2 THE DATA SETS

Four 'high quality' data sets were considered in the modelling test. In view of the stepwise test procedure, only data sets were considered that contained data on soil hydrology, solute transport behaviour, soil temperature, and, of course, soil pesticide fate. Further, given the European dimension of the project, data sets from different edapho-climatic regions of Europe were considered. In addition, given the availability of models explicitly considering macroporous flow, data from a typical macroporous soil were included. Finally, to evaluate the modelling performance of the balance terms, and to test the applicability of the modelling methodology to describe field behaviour, both field and lysimeter data sets were included. The references to the four data sets are given in Table 3.

TABLE 3 Data sets considered in the modelling exercise

Data set	Type	Reference of data set description	Reference to major modelling results
Vredepeel	Field	[6]	[27], [24]
Weiherbach	Field/Lysimeter	[22]	[12]
Tor Mancina	Lysimeter	[10]	[9]
Brimstone	Field	[13]	[2]

2.3 THE IMPLEMENTATION

Three workshops were organised to implement the modelling test. In the first workshop (Limburgerhof, Germany, autumn 1995) the participants were introduced to the modelling protocol. Data sets were presented and distributed after the initial workshop. During the second workshop (Piacenza, Italy, summer 1996), an intermediate evaluation of the study was done and participants presented their results related to soil hydrology, solute and heat transport. In a final workshop (Louvain-la-Neuve, Belgium, spring 1997), the pesticide modelling results were presented and overall conclusions and recommendations were drawn on the basis of discussions in four subgroups. In between the workshops, considerable exchange of information between model users, model co-ordinators, data co-ordinators, model evaluators and the organising committee took place.

3. Major recommendations and conclusions

The study resulted in a range of recommendations, which are shortly resumed below. All detailed simulations and analysis are given in [30] and papers therein.

3.1 THE DATA SET

Notwithstanding an attempt was maid to develop a high quality validation data set, several drawbacks were formulated. Given the within field heterogeneity, the measurement (and validation) of fluxes at the field scale remain uncertain. In addition, no evaluation of the fate and transport modelling within the sub-surface vadose zone could be performed. The following recommendations were formulated :

Related to the preparation of the data set:

- Characterise the history of the field site; and
- Design the experiment numerically.

Related to the boundary condition definitions of the experimental data set:

- Characterise high intensity rainfall better;
- Characterise actual evapotranspiration better;
- Characterise fertiliser and pesticide input better;
- Characterise runoff and volatilisation losses better; and
- Characterise lower boundary condition.

Related to the spatio-temporal resolution of the adopted sampling scheme:

- Use the soil horizons as carrier of information;
- Improve the description of the horizontal variability; and
- Improve the temporal resolution of the sampling.

Related to the analysed variables:

- Measure soil moisture content and tension;
- Measure soil temperature;
- Improve the inert tracer studies;
- Measures parent product and metabolites;
- Measure root densities and leaf area indices; and
- Characterise the agronomic practices. better

Related to the interpretation of the data set:

- Report raw data and give guidance on the estimation of model input from the data set;

- Report, if possible, field and laboratory data; and
- Pay particular attention to enhanced degradation, the selection of the moisture and temperature reduction functions, the selection of the sorption isotherm.

3.2 THE PHYSICAL TRANSPORT PROCESSES IN LEACHING MODELS

Each leaching model considers the modelling of the governing soil physical transport processes (soil moisture transport, soil solute transport, and soil heat transport). It was concluded that soil hydrological processes are important, but less sensitive than sorption and transformation processes. It was stated that hydrological models based on Richards equation are superior to capacity type models, if groundwater tables are close to the soil surface. The solute transport processes are not well understood and no agreement exists on the impact of hydrodynamic dispersion on pesticide leaching. Heat transport processes modelling was not felt to be a serious drawback in actual leaching modelling. It was felt that the temperature reduction mechanisms on pesticide degradation are far more uncertain than the temperature simulation itself. The following recommendations were formulated:

- Macroporous flow should be characterised better and work is needed for developing generic parameter estimation methods for macroporous flow models; and
- Inert tracer data should be made better available to improve the validation of the solute transport component.

3.3 THE SORPTION AND TRANSFORMATION PROCESSES IN LEACHING MODELS

Sorption and degradation submodels are also essential in leaching models. Enhanced sorption will decrease the mobility of the molecule in the soil root zone, while transformation processes reduce the total of chemical, which may leach to the subsurface groundwater body. The following recommendations were formulated related to the transformation and sorption models in current leaching models:
More accurate information on the impact of temperature and moisture on transformation constants is needed;

- Sub-soil degradation should be characterised better;
- More information on volatilisation and plant uptake is needed;
- More information of complex transformation kinetics is needed;
- More attention should be paid to the fate of metabolites;
- More information on the impact of hysteresis; and
- More information of the enhanced sorption is required.

4. The role of the model user in the pesticide leaching modelling

Leaching models are complex tools and often difficult to use correctly. Given the range of sensitive model parameters and model input which need to be specified, significant different modelling results may be obtained by different model results, even if similar original data sets are provided. Parameter estimation is therefore crucial in pesticide leaching modelling and a plea was made for the development of Good Modelling Practices (GMPs). The model user has its responsibilities. He needs to understand the model and its appropriate use, he needs to estimate appropriately the parameters, and keep in touch with new versions. The following recommendations were made:

- Evaluate the robustness of actual model input and parameter estimation procedures and implement robust procedures in practical guidelines; and
- Promote Good Modelling Practice, a.o. by the organisation of adequate training in leaching modelling.

5. Conclusions

In this paper, major recommendations from a recent European experience in pesticide leaching modelling are summarised. Twelve pesticide leaching models were tested by 36 model users, using experimental data collected at 4 different sites. An attempt was made to qualify both the models and model users. To our knowledge, this test is the most extensive one that has been carried out with pesticide leaching models in Europe so far.

It was confirmed that mathematical modelling offers major advantages for identifying the leaching of pesticides as compared to laboratory and field studies, yet no ' universal ' leaching model exists. It was concluded that more studies are needed to improve the validation status of leaching modelling. Within these studies, not only model validation should be considered but also modelling validation. This implies that the robustness of guidelines and parameter estimation procedures should be evaluated and that good modelling practice should be promoted. Given the quality of data that is needed for realising a validation study, network research should be promoted. It was finally felt that process descriptions in current models could be improved and that particular attention should be devoted to non-equilibrium phenomena in the description of the transport, transformation and sorption processes such as macropore flow in the physical transport, kinetic adsorption and time dependent degradation.

6. Acknowledgement

The results presented could only be realised through a network research sponsored by different local organisations and the European Commission through the Co-operation on Science and Technology Programme, COST66: 'Pesticides and the environment'. The contributors (K. Aden, A. Armstrong, N Baran, I. Bärlund, J. Boesten, D. Brockie, R. Bromilow, C. Brown , E. Capri, J. Catt, B. Diekkrüger, S. Ducheyne, M. Dust, O.M. Eklo, B. Erzgräber, G. Errera, R. Francaviglia, G. Görlitz, B. Gottesbüren, V. Gouy, E. Granitza, G. Harris, H. Hosang, J. Hutson, N. Jarvis, M. Klein, K. Kolle, H. Leeds-Harrison, A. Matthews, C. Mouvet, P. Nicholls, A. Portwood, S. Rekolainen, B. Remy, H. Resseler, H. Schäfer, P. Sweeney, A. Tiktak, M. Trevisan, T. van der Linden, L. van der Pas, H. Vereecken, A. Vischetti, and A. Walker) are firmly acknowledged for performing the simulation studies.

7. References

1. Aden K. and B. Diekkrüger, (2000). Modelling pesticide dynamics of four different sites using the model system SIMULAT. *Agricultural Water Management*, **44**, 337-357.

2. Armstrong A., K. Aden, N. Amraoui, B. Diekkrüger, N. Jarvis, C. Mouvet, P. Nicholls and C. Wittwer, (2000a). Comparison of pesticide leaching models: results using the Brimstone Farm data set. *Agricultural Water Management*, **44**, 85-105.

3. Armstrong A.C., A.M. Matthews, A.M. Portwood, N.J. Jarvis and P.B. Leeds-Harrison, (2000b). CRACK-NP: A pesticide leaching model for cracking clay soils.*Agricultural Water Management*, **44**, 183-200.

4. Boesten J.J.T.I and B. Gottesbüren, (2000). Testing PESTLA by two modellers for bentazone and ethoprophos in a sandy soil. *Agricultural Water Management*, **44**, 283-306.

5. Boesten J.J.T.I, M. Businelli, A. Delmas, V. Edwards, A. Helweg, R. Jones, M. Klein, R. Kloskowski, R. Layton, S. Marcher, H. Schäfer, L. Smeets, M. Styczen, M. Russell, K. Travis, Walker and D. Yon, (1995). Leaching models and EU registration. The final report of the work of the Regulatory Modelling Work Group of FOCUS, Forum for the Co-ordination of pesticide fate models and their use. Guidance document 4952/VI/95, Commission of the European Communities, DGVI, Brussels, 123 pp.

6. Boesten J.J.T.I. and L.J.T. van der Pas, (2000). Movement of water bromide ion and the pesticides ethoprophos and bentazone in a sandy soil: description of the Vredepeel dataset.*Agricultural Water Management*, **44**, 21-42

7. Camase, Concerted Action for the development and testing of quantitative Methods for research on Agricultural Systems and the Environment (1996). Register of agro-ecosystems models. DLO Research

Institute for Agrobiology and Soil Fertility, Wageningen, the Netherlands. http://www.bib.wau.nl/camase.

8. Dust M., N. Baran, G. Errera, J.L. Hutson, C. Mouvet, H. Schäfer, H. Vereecken and A. Walker, (2000). Simulation of water and solute transport in field soils with the LEACHP model,*Agricultural Water Management*, **44**, 225-246.

9. Francaviglia R., E. Capri, M. Klein, J. Hosang, K. Aden, M. Trevisan and G. Errera, (2000). Comparing and evaluating pesticide leaching models. Results for the Tor Mancina data set (Italy).*Agricultural Water Management*, **44**, 135-152

10. Francaviglia, R. and E. Capri, (2000). Lysimeter experiments in Tor Mancina (Italy)*Agricultural Water Management*, **44**, 63-74.

11. Freijer J.I., A; Tiktak, S.M. Hassanizadeh and A.M.A. van der Linden, (1996) PESTRAS v3.1: A one-dimensional model for the assessment of the fate of pesticides in soil. Report 715501007. National Institute of Public Health and the Environment, Bilthoven, the Netherlands. 130 pp.

12. Gottesbüren B., K. Aden, I. Bärlund, C. Brown, M. Dust, G. Görlitz, N. Jarvis, S. Rekolainen' H. Schäfer, (2000). Comparison of pesticide leaching models: results using the Weiherbach data set. *Agricultural Water Management*, **44**, 153-182

13. Harris G.L., J.A. Catt, R.H. Bromilow and A.C. Armstrong, (2000). Evaluating pesticide leaching models: The Brimstone farm data set. *Agricultural Water Management*, **44**, 75-84.

14. Jarvis N., C. Brown and E. Granitza, (2000). Sources of error in model predictions of pesticide leaching: a case study using the MACRO model, *Agricultural Water Management*, **44**, 247-262

15. Jarvis N.J., L.F. Bersgtröm and C.D. Brown, (1995). Pesticide leaching models and their use for management purposes. In: T.R. Roberts and P.C. Kearny (eds.) *Environmental behaviour of agrochemicals.* John Wiley and Sons Ltd 196-220.

16. Jury W.A. and R. Ghodrati, (1989). Overview of organic chemical environmental fate and transport Modeling approaches. In: 'Reactions and movement of organic chemicals in soils'*Soil Science Society of America,* Special publication 22, 271-304.

17. Klein M., J. Hosang, H. Schäfer, B. Erzgräber and H. Resseler, (2000). Comparing and evaluating pesticide leaching models. Results of simulations with PELMO.*Agricultural Water Management*, **44**, 263-282

18. Loague K., R.S. Yost, R.E. Green, T.C. Liang, 1989. Uncertainty in a pesticide leaching assessment for Hawai. Journal of Contaminant Hydrology, **4**, 139-161.

19. Nash J.E. and J.V. Sutcliffe, 1970. River flow forecasting through conceptual models. 1. Discussion of principles. Journal of Hydrology, **10**, 282-290.

20. Nicholls P.H., G.L Harris and D. Brockie, (2000). Simulation of pesticide leaching at Vredepeel and Brimstone farm using the macropore model PLM. *Agricultural Water Management*, **44**, 307-316

21. Rekolainen S., V. Gouy, R. Francaviglia, O.M Eklo and I. Bärlund, (2000). Simulation of soil water bromide and pesticide behaviour in soil with the GLEAMS model.*Agricultural Water Management*, **44**, 201-225

22. Schierholz, I., H. Shäfer, K. Kolle, (2000). The Weiherbach data set - an experimental data set for pesticide model testing on the field scale. *Agricultural Water Management*. **44**, 43-62.

23. Thorsen, M., J. Feyen and M. Styczen, (1996). Agrochemical modelling. In :*Distributed hydrological modelling*. M.B. Abbott and J.C. Refsgaard (eds.) Kluwer Academic Publishers. p 121-141.

24. Tiktak A., (2000). Application of nine pesticide leaching models to the Vredepeel dataset. II Pesticide fate. *Agricultural Water Management*, **44**

25. Trevisan M., G. Errera, G. Goerlitz, B. Remy and P. Sweeney, (2000). PRZM-2 evaluation using Vredepeel data-set. *Agricultural Water Management*, **44**

26. Trevisan M., G. Errera, A. Vischetti and A. Walker, (2000). Varleach evaluation.*Agricultural Water Management*, **44**

27. Vanclooster M. and J.J.T.I. Boesten, (2000). Application of nine pesticide leaching models to the Vredepeel dataset. I Water and solute fate. *Agricultural Water Management*, **44**

28. Vanclooster M., S. Ducheyne, M. Dust and H. Vereecken, (2000). Simulating pesticide dynamics with the WAVE-model. *Agricultural Water Management*, **44**

29. Vanclooster M., J. Boesten and M. Trevisan, (2000). Pesticide leaching modelling validation. *Agricultural Water Management*. **44**.

30. Vanclooster M., J.J.T.I. Boesten and M. Trevisan, (2000). Mathematical modelling for assessing pesticide leaching in agricultural soils at the field scale. In: P. Jamet and J. Cornejo (eds.).*Methods in pesticide science*. To be published.

EVAPORATION OF PURE LIQUIDS FROM OPEN SURFACES

FREDRIC C. ARNOLD & ALFRED J. ENGEL
U.S. Environmental Protection Agency,
Washington DC 20460
USA

Abstract

We developed an estimation technique for the evaporation rate of volatile liquids from open surfaces in the presence of a flowing air stream which, we believe, is an improvement over previous methods, particularly when the air velocity is greater than about 50 centimeters per second (100 feet per minute). The estimation technique is predicated on solution of the classical mass transfer coefficient model with the gas-phase mass transfer coefficient estimated by the correlation of Mackay and Matsugu. The new technique requires only the vapor pressure and molecular weight of the vaporizing substance, the ambient air temperature and velocity, and the area of the liquid pool for its solution. Other properties of the ambient air are incorporated into the model by the kinetic theory of gases. Our model was tested against experimental results on 19 different organic compounds generated by four different experimenters over a wide range of experimental conditions. The model matched the experimental data well, usually within 20 percent. The model can be used to obtain estimates of the evaporation rates of organic liquids resulting from spills or open surface operations.

1. Introduction

When reviewing premanufacture notifications for the production of new chemical substances as mandated by the Toxic Substances Control Act, the U.S. Environmental Protection Agency must evaluate potential exposures of workers to new chemicals during their manufacture, processing, and use. Therefore, the Agency is interested in developing methods for predicting worker exposures from the minimal amount of physical property data available on new chemical substances. For those chemicals, which are sufficiently volatile to result in exposure by inhalation, the rate of evaporation into the ambient atmosphere must be predicted with reasonable accuracy.

J.B.H.J. Linders (ed.), Modelling of Environmental Chemical Exposure and Risk, 61–71.

To calculate the rate of evaporation of a liquid, not only the physical properties of the liquid, but also the properties of the ambient atmosphere such as the air temperature, pressure, and velocity must be known. Precise calculation of evaporation rates requires a detailed understanding of the energy transfer between the volatile liquid, the air, and any solid surfaces in contact with the liquid. It is virtually impossible to know or be able to measure all these conditions. However, if some simplifying assumptions can be made with regard to this complex system, it is possible to arrive at estimates of vaporization rates for a variety of volatile substances that are, in most cases, within twenty percent of experimental results.

The intent of this work was to develop a model of pure liquid evaporation from open surfaces based on the mass transfer coefficient correlation of Mackay and Matsugu [8] and compare the model to penetration and mass transfer coefficient models from the literature. The assumptions made in developing these models are the same. They include the following: the contaminant is emitted from a single source; the surface area of the liquid pool does not change appreciably with time; air velocity above the pool is constant; there is no back pressure from the vaporized chemical which retards the rate of evaporation; and, the ideal gas law applies to all constituents. The resulting models require only the molecular weight and vapor pressure of the contaminant, ambient pressure and temperature, liquid pool size, and local air velocity for their solution.
Derivation of the mass transfer coefficient model incorporating the correlation of Mackay and Matsugu is described in the following section. This is followed by a description of the penetration model and mass transfer coefficient models from the literature. Subsequent sections compare the predictions of evaporation rates using the models to experimental data.

2. The Mass Transfer Coefficient Model

Colburn and Drew [3] analogized mass transfer to heat transfer, which resulted in an equation for mass transfer across an interface:

$$N_A = K \frac{p_v}{RT} \tag{1}$$

where N_A = mass transfer rate, g-moles/cm^2-sec
K = mass transfer coefficient, cm/sec
p_v = vapour pressure of the evaporating liquid, kPa
R = universal gas constant, 8.314 x 10^3 cm^3-kPa/g-mole-K
T = absolute temperature, K.

Mackay and Matsugu [8] used data on the evaporation of water and cumene from shallow open pans to correlate the gas-phase mass transfer coefficient as a function of wind-speed and liquid pool size by the equation:

$$K = 0.0220 \cdot v_z^{0.78} \cdot \Delta z^{-0.11} \cdot Sc^{-0.67} \tag{2}$$

where K = mass transfer coefficient, cm/s
v_z = wind speed, cm/s
Δz = length of the air-liquid interface in the direction of flow, cm
Sc = Schmidt number, the dimensionless ratio of air viscosity divided by the product of air density and diffusivity of the evaporating species, i.e. $Sc = \mu/\rho \cdot D_{AB}$.

The exponents on the velocity and length terms result from the Von Karman similarity principle as modified by Sutton (1953) and are reported to have provided satisfactory results when the airflow is in or near the turbulent range. The Schmidt number is used to correlate mass and momentum transfer between dynamically similar systems. The transfer rate can be converted to mass units by the relationship:

$$G = N_A \cdot M_A \tag{3}$$

where G = liquid evaporation rate, g/sec/cm^2
M_A = molecular weight, g/g-mole.

Combining Equations 1, 2, and 3, we obtain:

$$G = 2.65 \cdot 10^{-6} \cdot \frac{M_A \cdot p_v \cdot v_z^{0.78}}{\Delta z^{0.11} \cdot Sc^{0.67} \cdot T} \tag{4}$$

The Chapman-Enskog Theory can be used to estimate the viscosity of a gas, in this case, air. Details of the approximation are given by Bird *et al.* [2]. Over the temperature range from 10 to 70 °C, the viscosity may be approximated by:

$$\mu = -9.426 \cdot 10^{-5} + 1.610 \cdot 10^{-5} \cdot \sqrt{T} \tag{5}$$

where μ = viscosity of air at the ambient temperature, g/cm/sec
T = absolute temperature, K.

Similarly, the Chapman-Enskog Theory can be used to estimate the diffusion coefficient, D_{AB} , if the absolute temperature, pressure, and the molecular weight of the evaporating species are known:

$$D_{AB} = \frac{4.09 \cdot 10^{-5} \cdot T^{1.9} \cdot \sqrt{\frac{1}{28.97} + \frac{1}{M_A}}}{P \cdot M_A^{\ 0.33}} \tag{6}$$

where D_{AB} = diffusivity of species A in air, cm/sec
P = ambient air pressure, atmospheres.

Finally, at pressures below 35 kPa, the density of air may be determined from the ideal gas law:

$$\rho = 0.352\frac{P}{T} \tag{7}$$

where ρ = density of ambient air, g/cm^3
T = ambient air temperature, K.

Combining Equations 5,6, and 7, an expression for the Schmidt number is obtained:

$$Sc = \frac{\mu}{\rho \cdot D_{AB}} = \frac{\left(-6.57 + 1.12 \cdot \sqrt{T}\right) \cdot M_A^{\ 0.33}}{T^{0.9} \cdot \sqrt{\frac{1}{28.97} + \frac{1}{M_A}}} \tag{8}$$

Although the Schmidt number is constrained by the foregoing assumptions, for general environmental assumptions, its value lies between one and two. Therefore, its impact on the mass transfer coefficient is relatively small. Equation 8 for the Schmidt number may be substituted into Equation 4 for the evaporation rate to yield a single equation for calculation of the rate of evaporation of an organic liquid from a quiescent pool:

$$G = 4.71 \cdot 10^{-6} \cdot \frac{M_A^{\ 0.78} \cdot p_v \cdot v_z^{\ 0.78} \cdot \left(1/28.97 + 1/M_A\right)^{0.25}}{\Delta z^{0.11} \cdot T^{0.4} \cdot \left(-5.87 + \sqrt{T}\right)^{2/3}} \tag{9}$$

3. Alternate Models from the Literature

Hummel, Braun, and Fehrenbacher [6] developed a penetration model for diffusion of an evaporating liquid into a flowing gas using Fick's second law of diffusion. In terms of the variables as defined above, their expression for the rate of evaporation is given by:

$$G = 8.68 \cdot 10^{-7} \cdot \frac{M_A^{0.833} \cdot p_v \cdot v_z^{0.5} \cdot (1/28.97 + 1/M_A)^{0.25}}{\Delta z \cdot T^{0.05}} \tag{10}$$

Lennert, Nielsen, and Breum [7] at the Danish National Institute of Occupational Health studied the performance of seven correlations for predicting evaporation rates. The models applicable to turbulent flow conditions ([5], [4] and [*] were tested in this study.

4. Evaluation of the Mass Transfer Coefficient Model

Several sources of experimental data were used to evaluate the validity of the mass transfer coefficient model. The single largest source of experimental measurements is a study by Pace Laboratories funded by the U.S. Environmental Protection Agency [12]. Additional data were obtained from the papers of Wade [13], O'Brien and Stutzman [9] and Gray [4]. All expressions for the rate of evaporation of a volatile liquid listed herein were compared to these experimental data.

4.1 PACE DATA

In this study, 14 organic liquids[1] were evaporated from a small, temperature-controlled pan set in a rectangular duct into an air stream passing over the pan[2]. More than 150 test runs were conducted over temperatures ranging from 5 °C to 50 °C and air flow rates between 50 and 750 cm/s. The predictions of the Penetration Model developed by Hummel *et al.* [6] and the Mass Transfer Coefficient Model developed herein are compared to the measured evaporation rates in Figure 1. Omitted from this graph are the measurements at 50 cm/s, which are given in Figure 5; at this velocity, the air flow is in transition from laminar to turbulent flow and the mass transfer coefficient model is inapplicable because it assumes fully developed turbulent flow. If the prediction methods were to match the experimental data perfectly, all of the data points would lie on the unit slope line on the graph. As can be seen, for turbulent flow conditions, the penetration model significantly over-predicts the evaporation rate; the average error is +30% and the standard deviation (the square root of the mean square error) is 49%. The mass transfer coefficient model overpredicts the experimental data slightly; the average error is +10% while the standard deviation is 32%.

The correlations applicable to turbulent flow studied by Lennart *et al.* [7]) were also

---

1 Hexane, n-heptane, octane, benzene, toluene, xylene, acetone, methyl ethyl ketone, octanone, methanol, propanol, pentanol 1-hexanol, heptanol, and 2-octanol

2 Although the Pace experiments included vaporization of water, the models discussed in this article are not recommended to predict evaporation rates for water.

compared to the experimental Pace data. Of these models, the NiOH-2 model provided the best match with an average error of -20% and a standard deviation of the error of 32%. The average errors of the BAU, SUBTEC, and Gray models were all near 70%.

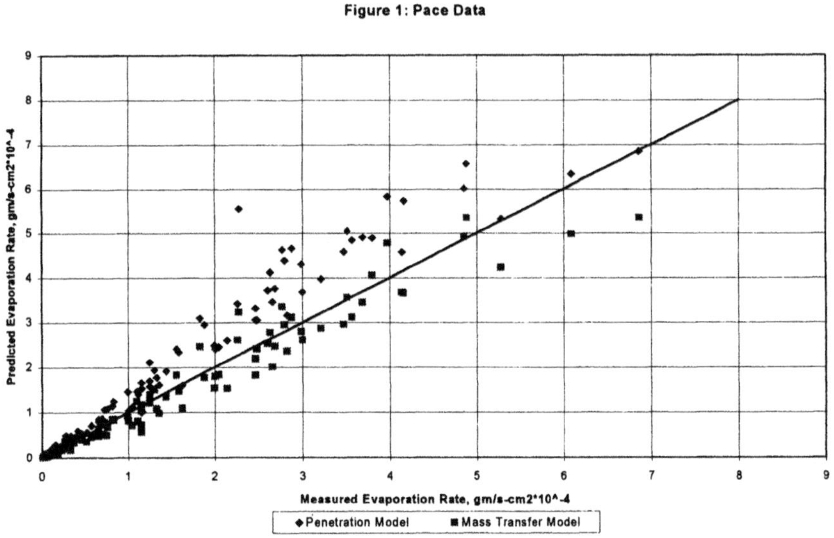

Figure 1. Comparison of the Penetration Model and the Mass Transfer Coefficient Model using the Pace Data.

4.2 DATA OF WADE

In 1942, Wade [13] reported his experiments on the rate of evaporation of six low-boiling liquids[3] from a shallow tray within a wind tunnel. His data was presented in graphical form, showing the measured evaporation rate as a function of vapor pressure with air velocity over the liquid pool as the parametric variable. Wade reported evaporation rates up to 0.15 grams per second and air velocities as great as 393 cm/sec (774 ft/min). The wind speed, liquid vapor pressure, and evaporation rate were read from the graphs and the liquid temperature inferred from the vapor pressure of the pure liquid. With this information, the predicted rate of evaporation could be calculated and compared to the experimental value.

Figure 2 compares the model predictions to the rates of evaporation measured by Wade. In general, the evaporation rates predicted with the Mass Transport Coefficient Model are superior to those predicted by the Penetration Model. The average error was -14% and the standard deviation was 4% using the former while the average error was 87% and the standard deviation 90 % using the latter. The NiOH-2 model of Lennart *et al.*

3 Acetone, benzene, carbon tetrachloride, ethyl acetate, toluene, and trichloroethylene

[7] provided estimates of the evaporation rate equivalent to the Mass Transfer Coefficient model but with a larger standard deviation.

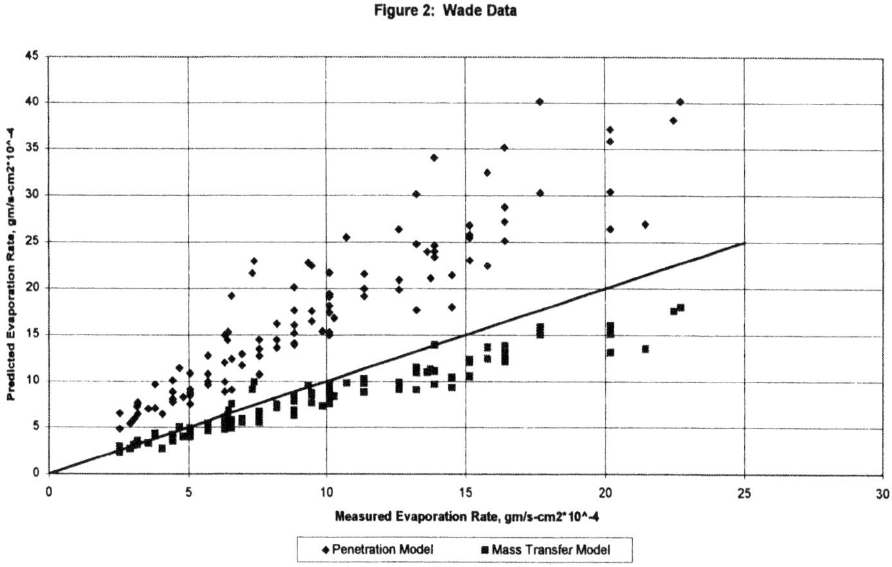

Figure 2. Comparison of the Penetration Model and the Mass Transfer Coefficient Model using the Wade Data.

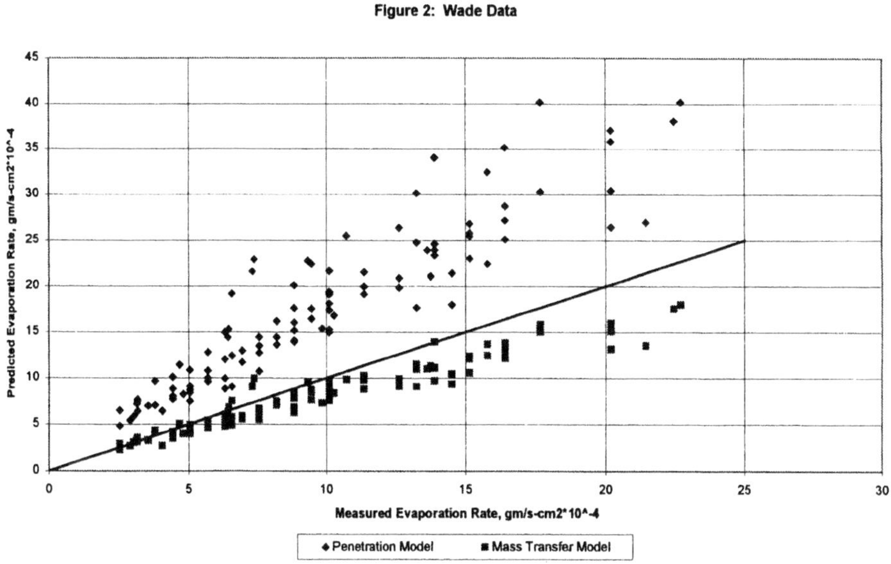

Figure 3. Comparison of the Penetration Model and the Mass Transfer Coefficient Model using the O'Brien Data.

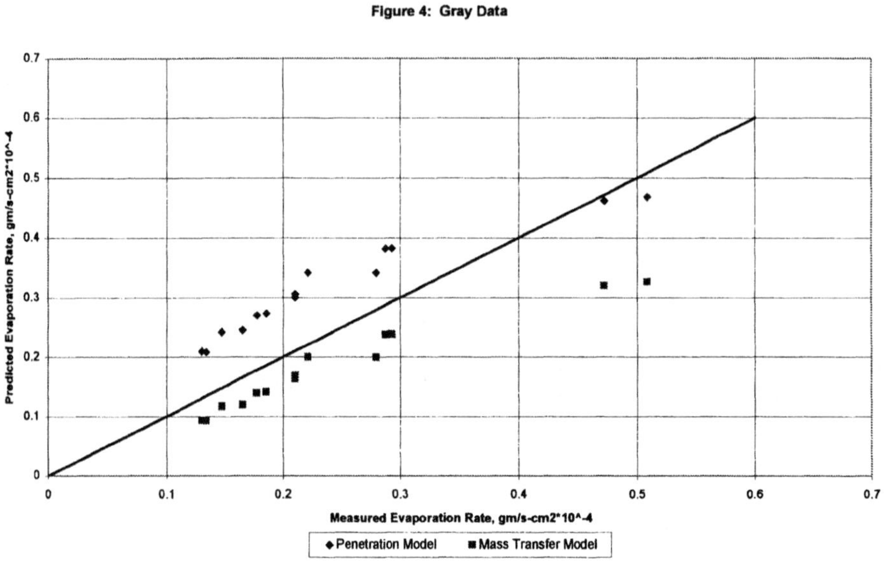

Figure 4. *Comparison of the Penetration Model and the Mass Transfer Coefficient Model using the Gray Data.*

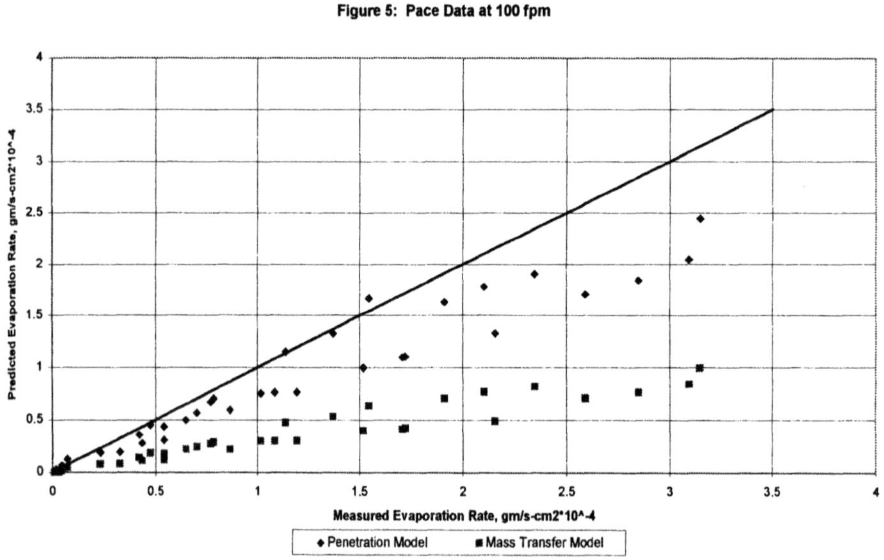

Figure 5. Comparison of the Penetration Model and the Mass Transfer Coefficient Model using the Pace Data at a speed of 100 feet per minute (50cm/s).

4.3 DATA OF O'BRIEN AND STUTZMAN

O'Brien and Stutzman [9] published a paper in 1950 in which they describe the evaporation of four volatile organic liquids in an apparatus similar to that used by Pace Laboratories and by Wade. In all of their runs, the air was heated to 38 °C while, in most cases, the liquid was also heated. Under these conditions, vapor pressures ranging from 3 to over 13 kPa were found to occur. Air velocities ranged from 0.3 to almost 2.5 m/s.

The results of applying the two models are shown in Figure 3. The Mass Transport Coefficient Model matches the data adequately with an average error of 13% and a standard deviation of 15%. The Penetration Model predicted the evaporation rates poorly. The average error of this model was 97% and the standard deviation was 101%. For the evaporation rates in this study, predictions made with the Mass Transfer Coefficient Model was superior for laminar as well as turbulent flow conditions. Of the models studied by Lennart and coworkers, the NIOH-2 model was equivalent to the Mass Transfer Coefficient model.

4.4 DATA OF GRAY

Douglas Gray reported data from the vaporization of xylene in a square wind tunnel, 15 cm on a side. The sampling well was 15 cm wide with a length of 15 cm and a depth of 0.64 cm. Air and liquid temperatures were between 21 and 25 °C in all experiments. Air velocity ranged from 11.9 cm/sec (23.4 ft/min) to 376 cm/sec (740 ft/min). The measured and predicted evaporation rates are compared in Figure 4. The Mass Transfer Coefficient Model tends to under-predict the evaporation rate for laminar flow conditions (when it is inappropriate) and to over-predict the evaporation rate when the air flow is turbulent. The average error of all predictions is -11% and the standard deviation is 17%. Predictions of evaporation rates made using the Penetration Rate Model are uniformly high; the average overestimation is 35% and the standard deviation of the errors is 42%. Of the models studied by Lennart and coworkers, the NIOH-2 model was comparable to the Mass Transfer Coefficient Model; the other models were less satisfactory.

5. Comparison at Laminar Flow Conditions

As noted previously, the comparison of predictive models for evaporation rates in this paper was intentionally limited to turbulent flow conditions at which the model of Mackay and Matsugu [8] was developed. However, a survey by Baldwin and Maynard [1] of the wind speed in 55 work areas covering a wide range of workplaces produced an arithmetic mean speed of 0.3 meters per second. The measurements were highly skewed with 85% of the individual measurements below the mean value. Accordingly, airflow in many indoor environments is laminar.

The Pace data taken at a wind speed of 100 feet per minute (50 cm/sec) is in the transition region from laminar to turbulent. At this wind speed, all of the models considered in this paper predicted an evaporation rate of the volatile liquid that was less than the measured rate. The Penetration Model provided the best estimates at this wind speed with an average underestimation of 20%. Predictions of evaporation rates made with the Mass Transfer Coefficient Model and the other models were, on average, less than half the measured evaporation rate.

6. Conclusion

Both the Mass Transfer Coefficient and the Penetration Models are completely general and can be applied to predict the vaporization of any pure liquid for which the molecular weight, vapor pressure, and liquid temperature are known. Experience suggests that the vapor pressure should be measured or calculated by a suitable correlation at the liquid temperature.

Before judging the validity of the models, some discussion of the data available for evaluation of the models is in order. First, the experimental design used by Pace Laboratories heated the liquid to minimize heat transfer effects on the rate of vaporization. It is doubtful if this aim was achieved. Second, their determination of the airflow regime, laminar or turbulent, was, at best, questionable given the design of the test section of their experimental apparatus.

Neither Wade [13], O'Brien and Stutzman [9], nor Gray [4] made any effort to avoid heat transfer effects in their experiments. However, in practical situations compensation for heat transfer effects is perhaps impossible and so we must settle for the liquid temperature as being the variable that can be measured and the best indicator of the vapor pressure near the air-liquid interface.

In light of the foregoing, care should be used in using the models described in this paper to predict evaporation rates. Nevertheless, the following generalizations can be made:

The Penetration Model

1. Is most appropriately used at very low air velocities (less than 50 centimeters per second), such as are encountered in unventilated interior environments, and
2. Should probably not be used when the vapor pressure of the evaporating liquid is in excess of 4 to 5.3 kPa.

The Mass Transfer Coefficient Model

1. Will provide satisfactory estimates of liquid evaporation rates under virtually all conditions except when the air flow rate 50 centimeters per second or less, and,

2. Will yield acceptable estimates of vaporization at high vapor pressures and turbulent airflow conditions.

7. References

1. Baldwin, P.E.J. and Maynard, A.D. (1998) A survey of wind speeds in indoor workplaces*Ann Occup. Hyg.* **5**, 303-313.

2. Bird, R.B., Stewart, W.E., and Lightfoot, E.N. (1960)*Transport Phenomena.* John Wiley & Sons, Inc., New York.

3. Colburn, A.P. and Drew, T.B. (1937) *Trans. A.I.Ch.E.* **33**, 197-212.

4. Gray D.C. (1974) Solvent evaporation rates. *Am. Ind. Hyg. Assn. J.* **35**, 695-710.

5. Gmeling, J., Weidlich,U., Lehmann, E., and Frohlich, N. (1989) A method for calculating airborne concentrations of substances when emitted from liquid product mixtures*Staub-Reinh. Luft.* 49, 227-230 (in German).

6. Hummel, A.A., Braun, K.O., and Fehrenbacher, M.C. (1996) Evaporation of a liquid in a flowing air stream. *Am. Ind. Hyg. Assn. J.* **57**, 519-525.

7. Lennert, A., Nielsen, F., and Breum, N.O. (1997) Evaluation of evaporation and concentration distribution models - a test chamber study. *Ann. Occup. Hyg.* **41**, 625-641.

8. Mackay, D. and Matsugu, R.S. (1973) Evaporation rates of liquid hydrocarbon spills on land and water. *Can. J. of Chem. Engr.* **51**, 434-439.

9. Nielsen, F., Olsen, E., and Fredenslund, F. (1995) Prediction of isothermal evaporation rates of pure volatile organic compounds in occupational environments - A theoretical approach based on laminar boundary layer theory. *Annals of Occupational Hygiene* **39**, 497-511.

10. Olsen, E., Olsen, I., Wallstrom, E. and Rassmussen. (1992) On the substitution of chemicals - use of the SUBFAC-indexfor volatile substances. *Ann. Occup. Hyg.* 36, 637-652.

11. O'Brien, L.J. and Stutzman, L.F. (1950) Mass transfer of pure liquids from a plane, free surface*Ind. and Engr. Chem.* **42**, 1181-1187.

12. Sutton, O.G. (1953) *Micrometeorology.* McGraw-Hill, New York.

13. U.S. Environmental Protection Agency (1992) Evaporation rate of volatile liquids, 2nd ed. by K.O. Braun and K.J. Kaplin (NTIS pub. No. PB 92-232-305). National Technical Information Service, Springfield, Virginia.

14. Wade, S.H. (1942) Evaporation of liquids in currents of air. *Trans. Inst. Chem. Engrs. (London)* **20**, 1-13.

APPLICATION OF USES FOR ESTIMATION OF PEC FOR PESTICIDES AND HAZARD ASSESSMENT FOR AQUATIC ENVIRONMENT

VESKA KAMBOUROVA & KOSTA VASSILEV
National Centre of Hygiene, Medical Ecology and Nutrition
15 "D.Nestorov" Str.,
1431 Sofia,
Bulgaria

Abstract

Some issues of the modelling investigation carried out within the frame of the Danube Pesticide Study are discussed. The Netherlands' Uniform System for Evaluation of Substances (USES, Version 1.0) has been applied in order to estimate the Predicted Environmental Concentrations (PECs) in surface water for set of pesticides mainly used in the Danube region.

Based on the obtained Hazard Quotients priority list have been performed according the potential of the pesticides to cause short- and long-term effects to the aquatic organisms. The results provide information for decision making support and elaboration of reduction programmes for aquatic environment.

1. Introduction

According to their designation for use in agriculture, pesticides are inevitably introduced in the environment and may constitute a risk for non-target organisms and different environmental media.

Probably the mostly discussed concern in recent years is the presence of pesticide residues in natural water as a media which quality is essential to maintain the sustainable aquatic life and as a source for drinking water.

As a result of increased public awareness of the undesirable consequences of pesticides scientific surveys have been carried out in order to provide early indications of the possible environmental effects of plant protection products, in both aspects - spatially and temporally.

The necessity to embrace the problem is imposed also by the regulatory measures. A leading principle in the EU Directive 91/414/EEC is to ensure a high standard of protection and to avoid unacceptable influence on the environment in general and

J.B.H.J. Linders (ed.), Modelling of Environmental Chemical Exposure and Risk, 73–78.

harmful effects on human or animal health or on the groundwater quality. The generation of Predicted Environmental Concentrations (PECs) of pesticides for four environmental compartments: soil, surface water, groundwater and air is part of the requirements of Annexes II and III in the same document [1].
Besides the widely accepted expert judgement of the results obtained from the available laboratory and field tests, diverse mathematical models have been elaborated to comply environmental and health issues associated with pesticide application. In this context mathematical modelling should not be viewed as an alternative but as a complementary assessment method, allowing more flexible use of the data derived from these studies [2]. The most widespread application of models relates to the estimation of environmental exposure data, respectively PECs to non-target areas and to the water compartment in particular.

2. USES 1.0 - Concept

This paper discusses a specific issue of the modelling investigation carried out within the frame of the Danube Pesticide Study [3,4], aimed to examine the hazard potential for the aquatic environment of a set of agricultural pesticides mainly used in the Danube region.
For this purpose the Netherlands' Uniform System for Evaluation of Substances (USES, Version 1.0) has been applied. USES is designed as a decision support system for rapid quantitative assessment of organic chemicals [5]. For this reason a predefined set of criteria, assumptions, estimation methods, system parameters and default values is introduced in the system and used in the calculations.
It incorporates average environmental characteristics of the Dutch situation and it should be pointed out that it does not allow a comprehensive or site-specific risk assessments.
An integral part of USES is the Evaluation System for Pesticides (ESPE.1) which permits a quantification of the general risk of agricultural pesticides [6]. One objective of the system is to evaluate the safe conditions for the aquatic environment, including surface water.
The evaluation approach in USES is based on a standard procedure for hazard assessment. The scheme involves exposure and effects assessment, which quantified endpoints are compared. The results of the evaluation are the hazard quotients. A hazard priority list of the substances under evaluation could be generated.
The pesticides introduction in the environment is presumed as a "pseudo point" emission source and the local emission is estimated, which is in close relation to the application mode of the compounds.
For the evaluation of the distribution to the aquatic environment the following mathematical models are incorporated: SLOOT.BOX model for pesticide spray drift, drainage model for pesticide, and PESTicide Leaching and soil Accumulation model (PESTLA).
Endpoints for the aquatic compartments are:

- Predicted concentrations in surface water, modelled as a ditch with defined parameters, surrounding the territory of pesticide application. Except the initial peak concentration, the concentrations over 4, 21 and 28 days, as well as the annual average concentration in the water phase are calculated.

- Predicted concentration in groundwater.

Acute toxicity data and information on no adverse effect levels for fish, crustaceans and algae are part of the input set and basis for extrapolation to derive no effect concentration (NEC) for the aquatic system as a whole. Extrapolation factors 10, 100 or 1000 are used, depending on the availability of experimental data.

Correspondingly two kinds of hazard quotients for aquatic organisms are the output of the evaluation module of USES.

Both of them have been used to assess the hazard potential of pesticides for the aquatic life:

- PEC/L(E)C_{50} ratio, which presents a direct comparison of PECs in surface water over 4 days with experimentally defined toxicity data L(E)C_{50} for each separate species, and
- PEC/NEC quotient for the whole ecosystem which reflects the ratio of the estimated exposure concentration (PEC) over the defined period of time and the level at which no adverse effects are to be expected for the most sensitive species. It is assumed that NEC ensures 95% protection of the species in the aquatic ecosystem [5].

3. Specific Case Application and Modelling Results

The study comprises 275 active ingredients. The data set filled in the system were formed based on the inventory of the pesticides used in the Danube region, the use and application patterns of the chemicals, the physico/chemical properties and available degradation parameters of the selected pesticides, the toxicity and ecotoxicity characteristics of the compounds [7].

The outputs of the USES evaluation module have been used to prioritise the pesticides under interest according to their hazard potential for the aquatic life, both: short- and long-term effects.

Table 1 presents the criteria applied to classify the pesticides under study according their potential to cause acute toxic effects on different aquatic species, used by Linders *et al.* [8].

TABLE 1. Classification criteria

Exposure/toxicity ratio	Classification
more/equal to 10	VL - very large
10 - 1	L - large
1 - 0.1	P - present
0.1 - 0.01	S - small
less 0.01	N - negligible

Totally 83 active ingredients have been classified as hazardous causing acute toxic effects on different aquatic trophic levels.

In Table 2 the distribution of the acute hazard potential for different aquatic species of the pesticides evaluated is summarised according to the applied classification.

TABLE 2. Distribution (%) of the estimated acute hazard potential of the herbicides evaluated

Species	Hazard category				
	VL	L	P	S	N
Algae	0	8.3	19.4	16.7	55.6
Crustaceans	0	0	0	1.0	99.0
Fish	0	0	0	4.5	95.7

Due to their nature, algae are most sensitive to herbicides. Approximately 50% of them has been classified in the hazard categories from "Large" to "Small". Monolinuron, Terbutryn, Metribuzin have been identified with the highest potential for acute effects. Taking into account the relatively low PECs estimated for the surface water these results can be associated with the high acute toxicity (EC_{50} range: 0.001 - 0.003 mg/l) of the compounds. Diquat, Dimefuron, Chlorbromuron, Methabenzthiazuron, Metoxuron, Metolachlor and Prometryne show "Present" hazard. Only Linuron has been evaluated to present small hazard for both species - crustaceans and fish. In addition Propachlor, Chloridazon and Alachlor have been classified in the same acute hazard category for fish. It might be summarised that with respect to their chemical nature the herbicides with the highest potential for acute damage of the aquatic organisms belong to the urea, triazine and chloracetanilide compounds.

The obtained results (Table 3) show that fungicides are relatively non-hazardous for

TABLE 3. Distribution (%) of the estimated acute hazard potential of the fungicides evaluated

Species	Hazard category				
	VL	L	P	S	N
Algae	0	3.3	0	0	96.7
Crustaceans	0	1.7	1.7	12.1	84.5
Fish	0	0	4.6	9.2	86.2

algae, except Fentin-acetate, which is in category "Large", with EC_{50} of 0.002 mg/l. Thiram and Pyrazophos present some hazard for Daphnids.

Fish seem to be sensitive to a relatively greater number of fungicides. About 13.8% from the evaluated active substances have shown the potential to provoke acute effects by the predicted concentrations, mainly in the range of 0.001 - 0.006 mg/l. Among them, in category "Present", are Dichlorfluanid, Fentin acetate and Folpet. Pyrazophos, Chlorothalonil, Ziram, Chinomethionat, Anilazine and Fentin hydroxide have been categorised in hazard group "Small". Nevertheless it should be mentioned, that most of the substances have acute toxicity parameters - LC_{50} (96 h) for fishes between 0.01 and 0.04 mg/l.

It is evident that insecticides are the group of pesticides most hazardous for aquatic life. A great amount of the evaluated insecticides have been considered to present a potential acute hazard for fish. Only Endosulfan and Lindane have been put in the category "Large", both of them with very low levels for acute fish toxicity. Dicofol,

Methidathion, Triazophos, Quinalphos, Fonofos, Carbosulfan, Fenvalerate, Chlorpyrifos, Propargite and Zeta-cypermethrin show "Present" hazard for fish. The acute hazard for crustaceans is more expressive. That might be associated also with the very low LC_{50} figures, despite the low PECs. The investigated active ingredients are relatively uniformly distributed among the range "Large - Small". In descending order they are arranged as follows: Quinalphos, Fonofos, Mevinphos, Fenvalerate, Malathion, Trichlorfon, Carbosulfan, Phoxim, Dicophol, Triazophos, Furathiocarb, Fenitrothion etc. Organophosphorous compounds, carbamates, pyrethroids and organochlorines are the most dangerous groups of insecticides for aquatic life.

TABLE 4. Distribution (%) of the estimated acute hazard potential of the insecticides evaluated

Species	Hazard category				
	VL	L	P	S	N
Algae	0	0	0	0	100.0
Crustaceans	1.2	12.8	16.3	17.4	52.3
Fish	0	2.1	10.5	18.9	46.2

On the basis of PEC/NEC quotients (HQ) pesticides priority list has been prepared in order to demonstrate their potential to create long-term effects on the aquatic ecosystem as a whole.

TABLE 5. Selected list of the Hazard Quotients (HQ) of the pesticides for the aquatic ecosystem

common name	HQ	common name	HQ	common name	HQ
HERBICIDES		**FUNGICIDES**		**INSECTICIDES**	
Monolinuron	952	Dichlofluanid	479	Quinalphos	18400
Diquat	514	Thiram	216	Dichlorvos	6176
Terbutryn	130	Folpet	150	Mevinphos	5409
Metribuzin	102	Fentin-acetate	133	Fenvalerate	3800
Propachlor	90	Pyrazophos	95	Malathion	2530
Chloridazon	57	Chlorothalonil	86	Carbosulfan	2110
Dimefuron	36	Fenarimol	39	Endosulfan	1730
Atrazine	34	Fentin hydroxide	37	Phoxim	1629
Prosulfocarb	33	Chinomethionat	17	Lindane	1580
Chlorbromuron	28	Carbendazim	13	Furathiocarb	1390
Linuron	24			Heptenophos	922
Methabenz-thiazuron	22			Carbaryl	886
Acetochlor	18			Fonofos	602
Metoxuron	17			Zeta-cypermethrin	534
Simazine	16			Phosmet	498
Metolachlor	13			Parathion-methyl	473
Prometryne	12			Carbofuran	419

A selected list of the pesticides with highest Hazard Quotients (HQ) is presented in Table 5, indicating the potential of each single pesticide. The hazard assessment ranges totally 40 herbicides, 19 fungicides and 65 insecticides in the group of those dangerous for the aquatic life, with a hazard ratio more than 1. Presented in percentage of the total number of evaluated pesticides it corresponds to 34.8%, 29.2% and 66.3 % respectively.

4. Conclusions

The presented results provide the basis to confirm the appropriateness of the model USES for screening comparative hazard assessment of pesticides for different aquatic species.

The likely response of the aquatic ecosystem as a whole might be predicted and on the basis of the derived Hazard Quotients an indicative priority list for regulatory measures might be generated.

The general evaluation of the hazard extent to non-target aquatic organisms may serve as useful information for initiating more refined research, decision making support and elaboration of reduction programmes for the aquatic environment.

It should be stressed that the relevance of the results depends on the adequateness of the input data and their completeness, as well as on the assumptions made to determine the more realistic worst case situation.

5. References

1. 91/414/EEC Council Directive concerning the placing of plant protection products on the market. (1991) *Official Journal of the European Communities* **L** 230,1-14.

2. Mackay, N. (1996) Proposal: Agrochemical Risk Assessment Dialogue Group for Europe, Proposal to SETAC - Europe.

3. Kambourova, V., Simeonov L., Dura Gy., Vassilev K.,Tasheva M. (1998) Comparative Hazard Assessment of pesticides to aquatic life using estimated concentrations. *Central European Journal of Occupational and Environmental Medicine* **4**, 343-353.

4. Dura Gy., Kambourova V., Vassilev K., Tasheva M., Simeonov L. (1998) Estimating Environmental Concentrations of Pesticides and Hazard Categories using Exposure Modelling. *Central European Journal of Occupational and Environmental Medicine* **4**, 328-342.

5. USES (1994) *Uniform System for the Evaluation of Substances (USES), version 1.0.* RIVM, VROM, WVC. The Hague, Ministry of Housing, Physical Planning and Environment. Distribution no. 11144/150.

6. Linders, J.B.H.J., and Luttik, R. (1995) Uniform System for the Evaluation of Substances. V. ESPE, Risk Assessment for Pesticides. *Chemosphere* **31**, 3237-3248.

7. Tomlin, C. (eds.) (1994). *The Pesticide Manual Incorporating the Agrochemicals Handbook.* Tenth edition, The British Crop Protection Council, Farnham, Surrey.

8. Linders, J.B.H.J., Jansma, J.W., Mensink, B.J.W.G. and Otermann, K. (1994) *Pesticides: Benefication or Pandora's Box ? A Synopsis of the Environmental Aspects of 243 Pesticides.* Report No. 679101014. National Institute of Public Health and Environmental Protection. Bilthoven, the Netherlands.

MODELLING OF OPERATOR EXPOSURE

ANTONELLA FAIT and BENGT S. IVERSEN
International Centre for Pesticide Safety
Via Magenta, 25
20020 Busto Garolfo (MI)
Italy
tel. +39-0331406611

Abstract

The methodologies are reviewed for predicting exposure to pesticides of agricultural operators, in the context of pre-registration risk assessment, with specific focus on the European registration. Exposure modelling allows predictions of exposure doses, which may be used in risk assessment for decision-making. Relevant targets for risk assessment are mixers, loaders, applicators, workers and bystanders. Risk can be described as a ratio between predicted dose and acceptable dose (AOEL), within a three-tiered approach. Exposure determinants during mixing/loading and application activities are discussed in the text. The skin is the main route of uptake of pesticides. The levels of exposure are relatively independent on the specific chemical properties of the substance, while mostly vary with the physical type of the formulation, use pattern and degree of personal protection. Generic exposure databases are collections of exposure data developed to predict, through mathematical calculations, exposure levels for typical scenarios of exposure. Generic databases are discussed in the text, which were developed in North America (PHED), UK (POEM), Germany, The Netherlands. EUROPOEM is a new promising tool, under development in Europe, which will allow evaluations at the European scale, and subsetting of exposure scenarios typical of different countries.

1. Introduction

There are a number of different purposes and methodologies for modelling "operator exposure", including its use in risk assessment, epidemiology, and other fields of science. This presentation is intended to review the methodologies used for predicting exposure to pesticides of agricultural operators, in the frame of pre-registration risk assessment, with specific focus to the European registration. In this context, modelling of exposure allows predictions of exposure doses, which may be used in risk assessment for decision making.

J.B.H.J. Linders (ed.), Modelling of Environmental Chemical Exposure and Risk, 79–89.

A few introductory remarks are given here, to review the legislative framework, define the concepts of operator risk assessment and "acceptable exposure", and highlight the determinants of exposure which are relevant under varying exposure scenarios.
We will then provide an overview of existing models, with specific reference to the models extensively used in Europe, describe the role they have in the risk assessment procedure, illustrate the input data required and the exposure estimates they generate.

2. The European registration of pesticides

The European legislation concerning the authorisation of Plant Protection Products (PPP) on the market is harmonising data requirements and procedures for pesticide risk assessment within the European Union. Following Council Directive 91/414 [1], the registration of pesticides requires approval of active substances at the European level. A positive list of active substances is in preparation (Annex I), which at present includes only three active substances: imazalil, azoxystrobin, and kresoxim-methyl. Plant protection products are evaluated and authorised at the National level, according to harmonised criteria established in Annex VI to Council Directive, the so-called "Uniform Principles" [2].
Directive 91/414/EEC states that no authorisation for an active substance or a PPP should be granted unless it is established that, under the proposed conditions of use (dose, application method, climatic conditions, etc.), it has no harmful effect on man, animals and the environment. Annex VI lays down uniform principles concerning the evaluation: which environmental and health aspects have to be evaluated for pesticide registration, how they have to be evaluated and the criteria that have to be met. Member States shall make a judgement as to the likely risks to humans, animals and the environment. The assessment of the health risk arising from the plant protection product is therefore a major component of the decision process.

3. Risk assessment of pesticide operators

According to Council Directive 91/414, the relevant human targets for risk assessment are the general population, pesticide operators (including mixers, loaders and sprayers), workers (those who may later re-enter treated crops for harvesting, weeding, inspection, etc), and bystanders (those who may be incidentally exposed at the time of application).
For risk assessment of pesticide operators, the authorisation cannot be granted if the likely levels of exposure, according to the proposed conditions of use, under the assumption of Good Agricultural Practices, exceeds the so-called "Acceptable Operator Exposure Level" (AOEL).
Exposure and dose information are combined with dose-response relationships to estimate risk, the probability of an adverse effect occurring. As for any other chemical, in fact, risk assessment of pesticides involves two fundamental components:

- on one side, a detailed evaluation of the intrinsic toxicity of the active substance, that is of the inherent capability to cause adverse effects and of the relevant end-points; the dose-response evaluation process allows identifying a **No Observed Effect Level** (NOEL);

the other component needed is the assessment of **exposure** that is the prediction of doses to which relevant targets are exposed in different exposure scenarios. This involves the estimate or measurement of the amount of a compound that reaches the body, and of the amount absorbed through each potential route of absorption. Exposure to pesticides mainly occurs through the dermal route, while inhalation is generally less important.

From these two basic processes, risk can be described, for specific exposure scenarios, as a ratio between the predicted exposure (or dose) and the AOEL.

4. The tiered approach

A three-tiered approach for risk assessment of pesticide operators was proposed at international Workshops held in The Netherlands and Canada [3,4], and incorporated into an OECD Guidance Document [5]. The principle of the tiered approach (Figure 1) is to provide an evaluation scheme, which allows the investigator to choose the most appropriate means of estimating or measuring exposure, once an AOEL has been established. In **Tier 1**, the likely exposure which is compared with the AOEL, is based on modelling (Generic Exposure Databases) and assumptions for unknowns (extrapolation from animal to man, permeability of human skin, effectiveness of protective measures).

The principle of **Tier 2** is to refine the estimate of the absorbed dose, when the calculations in tier 1 indicate that the AOEL will be exceeded. Tier 2 is also based on generic databases, but the assumptions of tier 1 are replaced by actual data on skin penetration (measured in vivo in animal /man) and/or on protective clothing penetration (measured).

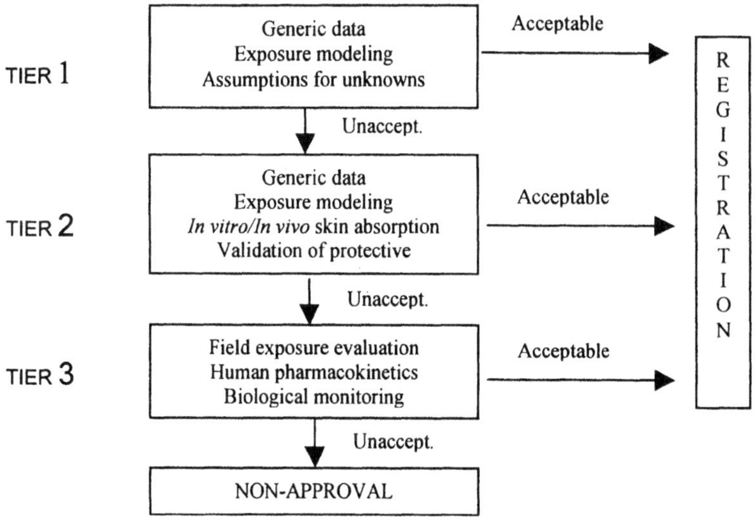

Figure 1. Tiered approach for risk assessment of pesticide operators [19].

Tier 3 is needed for use patterns where no generic data are available, or where the exposure estimated under tier 2 exceeds the AOEL. In these cases, field exposure studies, which may include biological monitoring data, must be considered to generate additional exposure data and improve the quality of risk assessment.

5. The concept of AOEL

The AOEL is defined as the maximum amount of active substance (a.s.) to which an operator may be exposed without adverse health effects. It can be defined as a systemic (i.e. internal) dose (mg a.s./kg b.w./day), expressed by the ratio "NOEL/SF", where:
- NOEL= No Observed Effect Level in the most sensitive adequate animal species, or in man when sufficient data are available
- SF= Safety Factor, accounting for uncertainties in extrapolation of toxicological data.

The process of AOEL setting requires the definition of the critical end-point and the identification of the NOEL in the most sensitive adequate animal species. The choice of the adequate experimental model strictly depends on the characteristics of anticipated exposure scenarios, and on considerations concerning extrapolation issues.

Results of laboratory studies in animals have to be extrapolated to man, to enable judgements to be made about the effects expected in humans. This is one of the most critical aspects of pre-registration risk assessment, and involves considerations about differences in routes of exposure, species, and individuals. Assessment factors are applied to the identified NOEL to account for uncertainties in the toxicological database, and in the extrapolation of data.

Specific criteria and procedures to establish AOELs for pesticides are currently debated within the European Union [6]. Several interpretations of the AOEL have been proposed, so far. An AOEL may be defined as an "external" dose, in which case two different limits are set (for dermal and respiratory exposure), and risk is estimated separately for individual exposure routes. An alternative approach is based on a single "internal" AOEL, in which case the health risk is evaluated as the ratio of the sum of exposure through individual routes, to the internal AOEL.

AOELs may be established for short-term and long-term exposure, depending on the anticipated exposure or use scenarios. A short- versus a long-term experimental model may be used as the source of the relevant NOEL. This has been proposed under the assumption that a short-term animal model better resembles the intermittent exposure of the operator, and provided it is demonstrated that the a.s. does not exert irreversible effects and does not accumulate in the body.

6. Occupational exposure to pesticides in agriculture

The term "external exposure" refers to the amount of a compound (a.s.) available for absorption through the oral, dermal, or inhalation route. It is generally assumed that the external exposure in the field mostly depends on the use scenario and not on the specific chemical substance used. In other words, the levels of external exposure are relatively independent of the chemical properties of the a.s., while mostly vary with the physical type of the formulation (granular, liquid, etc.), the use pattern (application technique and work method), and the degree of personal protection.

The levels of exposure depend on the working scenario [7,8]. The main working scenarios considered in Council Directive 91/414 are mixing, loading and application, all covered by the available exposure models. Other scenarios exist, e.g. re-entry exposure during harvesting, but modelling approaches for these scenarios are for the major part still under development.

A comprehensive overview of major exposure determinants during agricultural activities was published by Van Hemmen in 1993 [9], and some important aspects will be discussed here.

Mixing and loading of the PPP are two operations normally done sequentially by the same person, and therefore considered together, as a single scenario, in exposure models. Most PPPs have to be diluted from concentrated formulations, and mixed with water or other solvents either directly in the container used for application, or subsequently loaded into this. This operation with highly concentrated pesticides can give rise to serious exposure first of all on the operator's hands. The exposure during mixing/loading depends on the type of formulation (liquid or solid), the size of the solid particles, the size and shape of the container, the number of operations that have to be carried out, the amount of formulation used, as well as the loading technique employed.

The way the application takes place depends on the crops and area to be treated, and the available technique. The application can be done upwards (fruit trees) or downwards (horticulture, field crops), and manually or by vehicle-mounted equipment (in some cases, by aircraft). In greenhouses it is often done automatically. Exposure during application largely depends on the equipment and technique used, in particular on the particle size of the aerosol, and on the amount of pesticide applied, which depends on the area to be treated and on application time. During application, all body parts are at risk for exposure.

Environmental factors and also personal factors concerning the worker's way to do the job may further affect exposure. For example, wind speed and wind direction, or again, whether the work is done outdoor or indoor, are important determinants of exposure.

Large studies have been carried out to assess the exposure in the field covering different scenarios and varying climate conditions. Dermal exposure and inhalation are the two principal parameters for the assessment of the exposure and the risk, whereas ingestion normally is neglected. Only a part of the total amount of pesticide coming into contact with the protective clothing, work clothing and the skin, will be available for percutaneous absorption (actual dermal exposure), depending on the possible penetration of the garments worn. Also the amount inhaled depends on eventual protective equipment used. Inhalation is normally estimated by inhalation samplers, while dermal exposure is estimated by monitoring gloves, coverall or patches and undergarments. The skin is usually the main route of uptake of pesticides.

7. Exposure modelling

On the assumption that the spread of a cloud of aerosols is determined by environmental factors (such as climate), and agricultural factors (such as the particle size distribution of the aerosols, which depends on application equipment and volume rate), the level of exposure is largely determined by the job and the way it is done [9]. The selection of the "scenario" is therefore the first and most important step when assessing operator exposure for registration purposes.

The use of "generic databases" to predict operator exposure was first proposed in the mid 1980's [10,11] and exposure modelling for registration purposes is now widely accepted in Europe and North America.

"Generic databases" are collections of exposure data, representative of varying scenarios of pesticide use, which have been developed to predict exposure levels for typical scenarios of exposure. Fundamentally these models are databases with exposure data from selected and controlled field studies. These exposure data sets allow the prediction of surrogate exposure levels for typical scenarios of exposure. Surrogate data are derived from the database, depending on the scenario and pesticide formulation in question. Exposure is measured as the amount of a compound available for dermal and inhalation absorption.

Many generic databases are in use in different countries. Databases have been developed in North-America [12], in the UK [13,14,15], in Germany [16], and in the Netherlands [17]. Major differences among existing exposure models concern the source data sets, the scenario classification scheme, the considered type of dermal exposure (potential or actual), the statistics used to estimate exposure levels (geometric mean, 75th percentile, etc.), and the units used to express exposure measurements (volume of spray per unit of time, or weight of a.s. per amount of pesticide handled).

The Pesticide Handlers Exposure Database (PHED), developed by the US-EPA and Health Canada, is the predictive model used in North America. It is based only on North American field studies, and is fully computerised. The user is allowed to select very specific exposure subsets for analysis, such obtaining highly representative surrogate values. The PHED includes climatological and geographical conditions. The large existing differences in size of farms and agricultural methods between Europe and North America make the PHED model less suitable for European agricultural scenarios.

Exposure models, which are used extensively in Europe, are the German model [16] and the UK-Predictive Operator Exposure Model [15].

The exposure during mixing/loading depends on the type of formulation used (liquid, wettable powders or granules). The type of formulation is a model parameter in the German model and also in the UK-POEM model. The UK-POEM assumes that exposure during mixing depends only on the number of operations carried out with the container of the product. The UK model gives also nominal values of contamination per operation when emptying containers with varying volume and design.

The spray during application can be directed downwards or upwards, depending on the height and properties of the crop to be sprayed. Both models incorporate the choice between spraying of high or low crops. The upward spray greatly increases the risk of dermal and inhalation exposure.

The German model and the UK-POEM (see Figures below) do normally generate very different exposure estimates because they differ in:

- statistical approach to estimate surrogate exposure values: the German model estimates exposure level as the geometric mean of all individual data in its database, whereas the UK-POEM takes the 75th percentile of all individual data,
- different values of penetration factors for protective clothing (Table 1),
- different assumptions on the type of clothing worn by a non-protected operator,
- different correlation established between exposure and factors of influence.

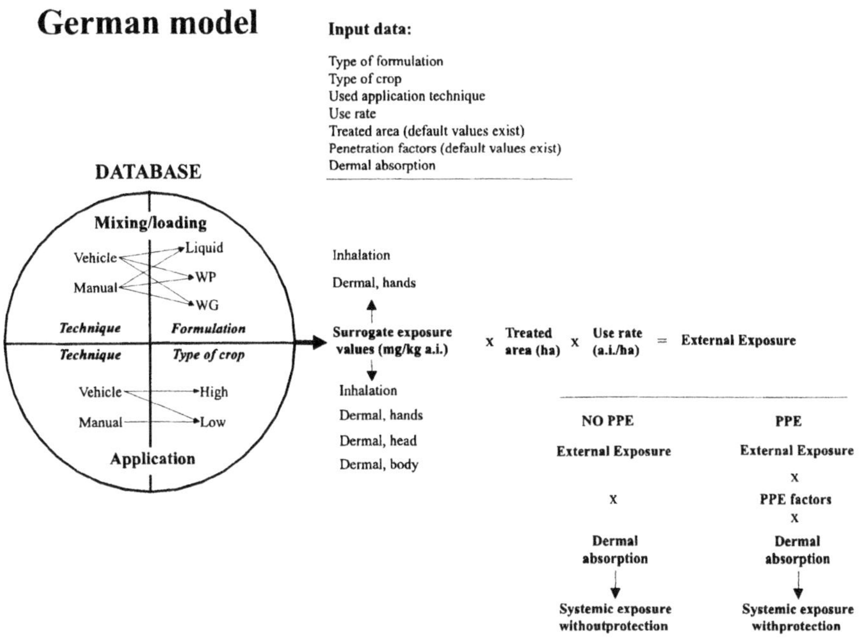

Figure 2. Exposure prediction, the German model.

Recently, an expert working group for the European Commission has developed a new database [18], which should be representative of exposure scenarios, at the European scale. It was in fact agreed that harmonisation of modelling was needed within the E.U. Different subsets of the database may be used to support the differences in climate and working practices in different Member States.
The underlying studies are mostly unpublished, and have been critically reviewed with respect to documented criteria [9]. The EUROPOEM [18] was developed to collect data that met standards of quality assurance and strict validation of the parent study. The standard quality criteria were: i) good documentation, describing in detail the study design, participants and results; ii) representativeness of the study for European situations; iii) adequate sampling methodology applied to different workers under varying conditions of equipment and climate, over a large part of the work shift; iv) adequate chemical analyses, with appropriate quality controls [19].

EUROPOEM includes data on a range of exposure scenarios, including mixing/loading of liquid formulations, wettable powders, wettable granules, and different application techniques using vehicle and hand-held methods, such as boom and pistol spraying, knapsack sprayers and airblast sprayers, upwards and downwards, indoor and outdoor.

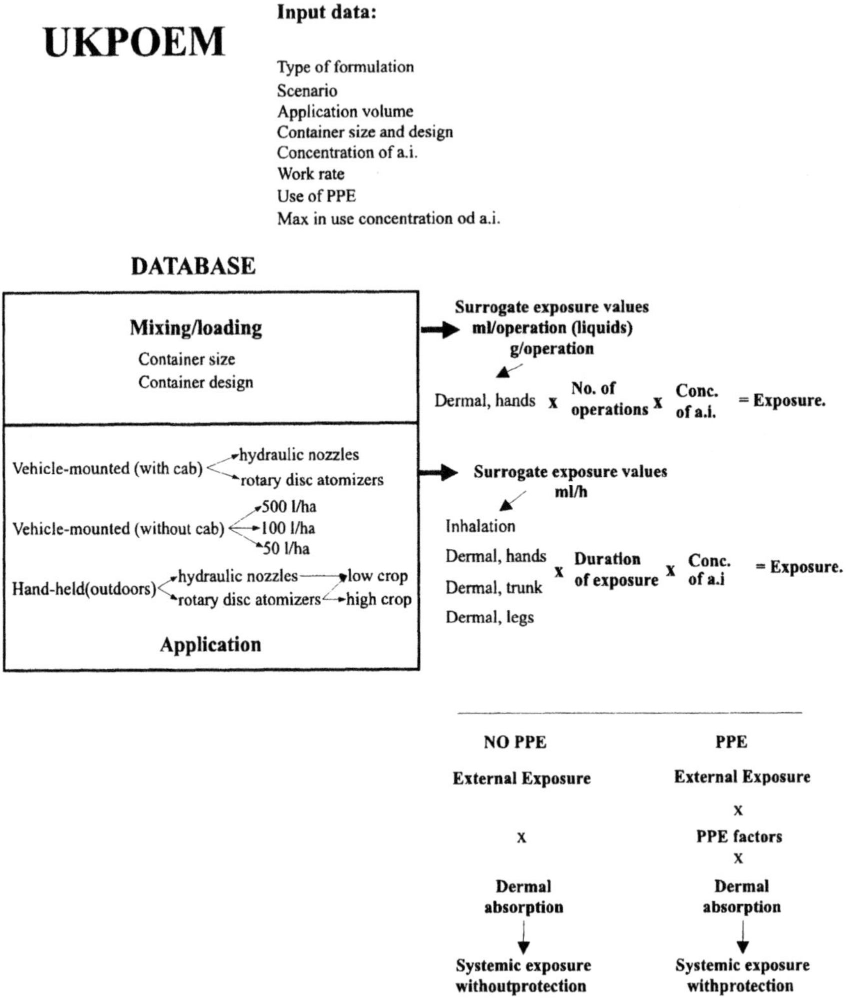

Figure 3. Exposure prediction, the UK-POEM model

Dermal and inhalation exposures are measured by patch techniques, whole body dosimetry, personal air pumps and fixed site air collectors. The model allows subsetting scenarios, making statistical analysis and exposure summaries. The first version of EUROPOEM contains about 40 studies with a total of about 750 data points. Still these 750 data points do not allow making estimates for all subsets of exposure scenarios, but these data gaps will be overcome in the next upgrades of the model.
The estimate of surrogate values is based on the 75th when at least 50-100 or more data points are available in the subset, from at least ten different studies. For subsets with 15-20 data points from three or less studies, no value is given, or the highest one. The 90th percentile is used in intermediate cases [19]. Exposure values are expressed as mg of

active substance per amount of pesticide handled (mg/kg), or as amount of formulation or spray (in mg or ml) per unit of time (ml/spray/hr).
One factor that should be taken into account when running exposure models is that the exposure reduction coefficients of gloves and other personal protective equipment vary between predictive models. TABLE 1 shows reduction coefficients used by the German model, the UK-POEM and the EUROPOEM.

TABLE 1 *Exposure reduction coefficients of personal protective equipment in different exposure models*

Protective clothing	German model	UK-POEM	EUROPOEM
Gloves	0.01		0.1
During mix./load. of solid formulations		0.01	
During mix./load. of SC's		0.05	
During mix./load. of EC's		0.1	
During spraying		0.1	
Normal clothing		variable	0.5
Coverall/overall with boots	0.05		0.1
Broad rimmed hat	0.5		
Face shield	0.05		
Half mask with filter	0.8		

8. Conclusions

The choice of surrogate generic values from Generic Databases is a crucial issue. The German model probably underestimates exposure, since it makes use of the Geometric Mean. The EUROPOEM is more conservative, since criteria were established to weight surrogate values, taking into account also the number of available studies and data points. The huge difference in estimated exposure levels calls for a very careful use of these models.
The accuracy of an estimated level of exposure depends, among other factors, on the amount of information available in the database. The specificity of different exposure scenarios is therefore limited in the models, because otherwise there would not be enough data to estimate an exposure level with sufficient statistical power.
One of the main limitations of existing Generic Databases is that none of them is representative of all exposure patterns typical of European countries. Within Europe, exposure scenarios largely vary in climate, working practices, cultural habits, etc. Generic Databases may predict exposure only under the assumption of Good Agricultural Practices, but the wide country-based variability in agricultural practices might bias the estimates. The expected bias is towards an underestimation of exposure, especially in Southern Mediterranean countries characterised by warm temperature, poor use of PPE, small crops, etc.
Ideally, different subsets of the database would be available to support these variations. In the EUROPOEM however, such a degree of differentiation would diminish the size of individual subsets so that insufficient data are available to select a surrogate value with confidence.

Existing models do not cover all exposure scenarios, and some special cases deserve further work. These are: greenhouse applications, seed treatments, grain storage, and granules/pellet applications.
The accuracy of exposure models depends entirely on the underlying databases. The majority of data comes from unpublished studies, covering only the most frequent scenarios. More good exposure studies are needed, and it is expected that the EUROPOEM model will be an important step forward in the prediction of exposure to pesticides. The development of EUROPOEM has already started, covering also exposure of bystanders and re-entry workers, and including the evaluation of mitigation measures.

9. References

1. CEC (1991) Council Directive 91/414/EEC of 15 July 1991 concerning the placing of plant protection products on the market; *OJ L230 19.08.91.*

2. CEC (1997) Council Directive 97/57/EC of 22 September 1997 establishing Annex VI to Directive 91/414/EEC concerning the placing of plant protection products on the market; *OJ L265 27/9/97.*

3. Henderson, P.Th., Brouwer, H.D., Opdam, G.J.J., Stevenson, H., and Stouten J. Th.J. (1993) Risk assessment for worker exposure to agricultural pesticides: Review of a workshop, *Annals of Occupational Hygiene*, .**37**, 499-507.

4. Carmichael, G.N. (1995) Pesticide Exposure. Overview of the Tier Approach. in: B.P Curry, S. Iyengar, A.P. Maloney and M. Maroni (eds),*Methods of pesticide exposure assessment,* Proceedings of a Workshop on Methods of Pesticide Exposure Assessment held in October 5-8, 1993, in Ottawa, Canada. Plenum Press, New York, pp.117-122.

5. OECD (1997) Guidance Document for the Conduct of Studies of Occupational Exposure to Pesticides During Agricultural Application, *Series on Testing and Assessment,* OECD Environmental Health and Safety Publications Paris, No.9.

6. CEC (1997) Guidance for the setting of AOELs. Draft Guidance Document 7531/VI/95, Directorate General for Agriculture. Commission of the European Communities.

7. Chester, G.(1993) Evaluation of agricultural worker exposure to, and absorption of, pesticides,*Annals of Occupational Hygiene*, **37**, 509-523.
8. Fenske, R.A. (1993) Dermal exposure assessment techniques,*Annals of Occupational Hygiene*, **37**, 687-706.

9. Van Hemmen, J.J.(1993) Predictive exposure modelling for pesticide registration purposes,*Annals of Occupational Hygiene*, **37**, 541-564.

10. Hackathorn, D.R., and Eberhart, D.C. (1985) Database proposal for use in predictive mixer-loader-applicator exposure, in R.C. Honeycutt, G. Zweig and N.N. Ragsdale (eds),*Dermal Exposure Related to Pesticide Use,* ACS Symposium Series, pp 341-355..

11. Honeycutt R.C. (1985): Field worker exposure: the usefulness of estimates based on generic data, in R.C. Honeycutt, G. Zweig and N.N. Ragsdale (eds),*Dermal Exposure Related to Pesticide Use,* ACS Symposium Series, pp 369-375.

12. PHED (1995) *Pesticide Exposure Handler Database, Reference Manual, Version 1.1.* Versar Inc. Springfield.

13. JMP (1986), *Estimation of exposure and absorption of pesticides by spray operators*, Joint Medical Panel of the Scientific Subcommittee on Pesticides of the Ministry of Agriculture, Fisheries and Food and the Toxicology Committee of the British Agrochemical Association. MAFF, Pesticides Registration Departement, Harpenden Laboratory, Herts.

14. Martin, A.D. (1990) A predictive model for the assessment of dermal exposure to pesticides, in R.C. Scott, R.H. Guy and J. Hadgraft (eds), *Predicition of Percutaneous Penetration, Methods, Measurements, Modelling* IBC Technical Services Ltd, Southampton, pp.273-278.

15. PSD (1992), *UK Predictive Operator Exposure Model (POEM). A Users Guide*. Pesticide Safety Directorate, Ministry of Agriculture, Fisheries and Food, York.

16. Lundehn, J.-R., Westphal, D., Kieczka, H., Krebs, B., Löcher-Boltz, S., Maasfeld, W., and Pick, E.D. (1992) Uniform principles for safeguarding the health of applicators of plant protection products. *Mitteilungen aus der Biologischen Bundesanstalt für Land und Forstwirtschaft*, Heft 277, Berlin.

17. Van Hemmen, J.J.(1992) Agricultural pesticide exposure data-bases for risk assessment, *Reviews of Environmental Contamination and Toxicology*, **126**, 1-85.

18. EUROPOEM (1997): *The Development, Maintenance and Dissemination of a European Predictive Operator Exposure Model (EUROPOEM) Database*, Final Report, BIBRA International, Carshalton, UK.

19. Van Hemmen, J.J.(1998) Pesticide worker exposure assessment for registration purposes. *Occupational Hygiene*, **4**, 243-257.

USE OF ALARM MODEL IN ACCIDENTAL POLLUTION OF DANUBE RIVER

Case study.

SILVIA CHITIMIEA & AUREL VARDUCA
National Institute of Research Development for Environmental Protection
Spl. Independentei No.294, sector 6, cod 77703,
Bucharest 78,
Romania

Abstract

The prediction of the propagation of a plume of polluted water along the river is considered to be an essential tool in a case of accidental pollution. Not only the predicted time of arrival of the polluted water should be the result of the calculation, but also its concentration. Prediction of the arrival time enables action to be taken at the right moment, being also used for estimating the impact of the pollution. For the Danube River and its tributaries a model – called DBAM - is needed like the Rhine Alarm model to estimate the propagation of a spill.

1. Introduction

The river Danube is formed by the confluence of two streams (Brigach and Breg) originating in Germany's Black Forest. The main river is 2857 km long and drains 817000 sq. km – a vast region that covers Hungary, Romania, Austria, Slovenia, Croatia and Slovak Republic as well as parts of Germany, Czech Republic, Bulgaria,.Moldavia and Ukraine. The catchment area also includes parts of former Yugoslavia and Bosnia, and smaller parts of Albania and Poland.

For Romania, the Danube stretch is of 1075 km, including the Danube Delta where the river discharges to the Black Sea.

The Danube Basin Alarm Model – DBAM -, like the Rhine Alarm Model, provides information of sufficient accuracy based on water level / flow – time relationships for various river sections, taking into account also the effects of stagnant zones and dispersion phenomena.

2. Background

Most of the accidents have impact on rivers (70%) followed by the lakes and groundwater.

J.B.H.J. Linders (ed.), Modelling of Environmental Chemical Exposure and Risk, 91–96.

The most frequent types of pollution were those due to oil residue (40-70%), pesticides and fertilisers.

After 1992, the last two categories decreased in the countries from Central and Eastern Europe.

Statistical data about causes of accidental pollution revealed that: 33% have unknown causes; 21% are due to technological accidents; 11% are due to unfavourable meteorological conditions; 16% are due to bad handling; 19% - others.

In **1980** danubian countries had become aware of need to meet together to protect the environment of the Danube. The first important environmental agreement is the **1985 Bucharest Declaration**.

In the period **1985-1987** a common programme was established and agreed for the trans-national Danube River monitoring network.

Starting from **1990**, in the frame of the Environment Programme for the Danube River Basin, two subgroups have been created:

- Monitoring, Laboratory and Information Management (MLIM), and,
- Accident Emergency Prevention and Warning System (AEPWS).

The AEPWS has for Danube River catchment area the following goals:

- ♦ to communicate the information about the sudden changes in water characteristics;
- ♦ to ensure an operative transboundary system for the main course of the Danube and for its tributaries.

3. Objectives

The main objectives may be distinguished in two groups:

a) Objectives of global importance:

- rapid communication of the information transmission about emergencies;
- reception, processing and prompt transmission of information about pollution, which may produce transboundary effect;
- to alert the responsible authorities in order to define the danger, establish the causes, determine the responsibility, remove the damages and prevent other consequences;
- to transmit the necessary information about the minor accident which may occur and may have possible impact on the pollution.

b) Objectives of local importance:

- to create an official framework able to prevent accidental pollution with a local origin;
- to prevent damages produced on water use and protect the downstream users;
- to prepare operative measures of action to limit the accidental pollution wave movement and also to limit the spreading area;
- to improve the framework and to harmonise it in a compatible structure with other bilateral or regional warning systems.

The data acquisition system is based on the facilities of the territorial laboratories (at the basic level), and the facilities of national laboratories, in special cases.

Generally, two information flows are defined: in vertical line from one level to another, and in horizontal line to the downstream units.
The DBAM forms an integrated part of the Danube Environmental Programme, being an operational model for:

- calculating the transport and decay of substances that have been released during accidental spills;
- estimating the propagation of an accidental pollution;
- knowing which water users of the Danube River are affected.

The system concerns the Danube course and the main tributaries as well.

4. Modelling concept

DBAM predict concentration levels of inadvertently released substances, as a function of location and time. This means that the model has to deal with a number of aspects. At first, different types of substances can be released in different ways. Secondly, the flow regime at the moment of spill should be characterised and incorporated into the model. Thirdly, there may be different pathways between the point of spill and the point(s) of observation. Finally, there is the model equation describing the behaviour of the substance(s) under consideration.
The basic assumption governing the modelling of the concentration is the advection-diffusion equation. And in addition, the common assumption of directional independence of the molecular diffusion and of incompressibility of water flow is used.
Furthermore an initial concentration at time "zero" is needed. For calamitous spill models this condition is generally zero. In case of an instantaneous spill, the initial concentration distribution at the moment of the spill can also be used.
All of these equations are associated with others for boundary conditions, for proportional decay of substances (for example for volatile substances only an additional term for the boundary condition at the water surface comes in), for existing turbulence, for spatially varying coefficients, etc.
The modelling concept is based on an analytical solution of the governing process that is applicable throughout the whole Danube Basin and for all the selected rivers in the basin.

5. Model simulation

5.1. COVERAGE OF THE RIVER BASIN

The model contains the Danube and all branches and tributaries that can be influenced by accidental spills from upstream industrial plants, industrial areas or shipping accidents. For those parts of the river basin where insufficient information is available, estimates might be inserted.
DBAM is constructed in such way that extension of the river schematisation is possible by adjustment of the input files only.

Data acquisition

Data, which has to be provided by the user of the model, should be as minimal and as unambiguous as possible and must consist of the same items with the same units and formats as the data provided by the AEPWS.
Data are referring to the time, location and duration of the spill, the amount of mass released and the decay coefficient of the substance.

5.2. MODEL OUTPUT

The simulation model consists of a flow-time module, evaluating the elapsed time from the spill to any location in the Danube River Basin on the basis of water level information and hydraulic characteristics, and of a dispersion module evaluation of the transport and spreading of spilt material.
Model simulation is able to cover the time span from the accidental spill to the time needed to flush the majority of the spilt material into the artificial lake of Iron Gate I or into the Black Sea.
DBAM conserves the mass for non-decayable substances and produces spatial distributions both as a function of the location along the axes of the main stream of the river basin network and as a function of the distance between the branches at any location along the axis. The output of then spatial distribution along the main axes is produced as a colouring of river stretches by an on-screen map and the output of spatial distribution between the river banks is produced as C-x (Figure 1) or C-y graph as a function of given step and at a selected location as a C-t graph (time function spatial concentration distribution) (Figure 2.). The computed results might be obtained also in a tabular form.

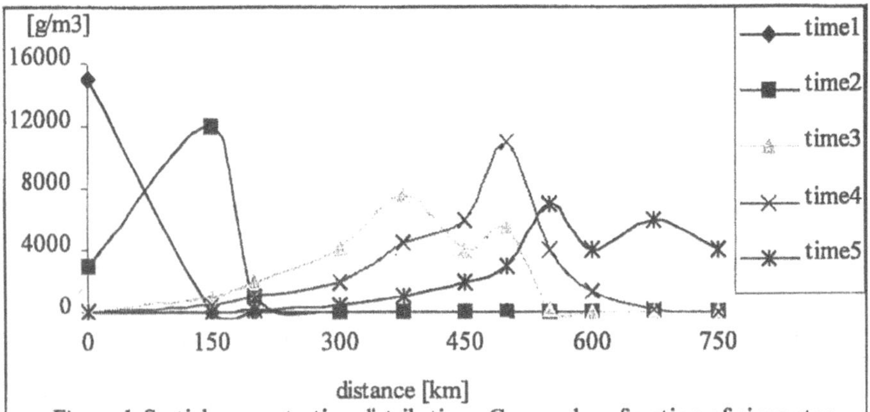

Figure 1. Spatial concentration distribution: C-x graph as function of given step

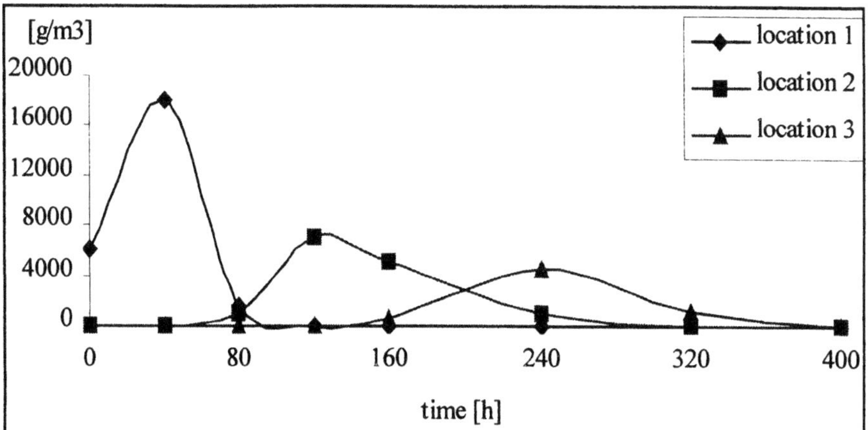

Figure 2. Spatial concentration distribution: C-t graph at selected locations

Model simulations produces laterally integrated results along the axis of the main stream of the river basin network, that show the skewness typical for instantaneous releases in river systems with so-called "dead zones".
The model results are reproducible, are produced with accuracy and are almost instantaneously shown.

6. Conclusions

Based upon the use of DBAM in simulation of cases of accidental pollution, the following conclusions can be underlined:

- the development of the model for the prediction of the propagation of a plume of polluted water along rivers in the Danube basin is feasible from the point of view of the modelling concept and from the point of view of data availability;
- the model developed is applicable for the upper Danube basin, the middle Danube basin and the lower Danube;
- the most important data needed to set-up the alarm model are present and can be made available;
- to have data easily available during the process of the model development, first calibration and operation of the model are required, in addition it is recommended that responsible authorities for the data be instructed from the highest possible authority level to provide the data for this purpose;
- the DBAM development allows for a stepwise upgrading and for geographical extension of the alarm model.

7. References

1. Environmental Programme for the Danube River Basin. Danube Basin Alarm Model Pre-study. Final report (1996), DELFT Hydraulics.

2. Mazijk, A. van, Mierlo J. van (1991) Several publications of the Faculty of Civil Engineering of the Technical University of Delft.

3. Somlyody, L. (1977) Dispersion measurement on the Danube, *Wat. Res.* **11**, 411-417.

4. The Set-up of the Danube Accident Emergency Warning System (1994), DELFT Hydraulics.

5. APELL - A Process for Responding to Technological Accidents (1988), UNEP.

RESULTS OF THE USE OF TWO ENVIRONMENTAL MODELS FOR PESTICIDES RANKING BY HAZARD

FINA KALOYANOVA[1], GYULA DURA[2], VESKA KAMBOUROVA[1]
[1]*National Centre for Hygiene, Medical Ecology and Nutrition, Bul. Dimitar Nestorov 15, Sofia 1431, Bulgaria*
[2]*National Institute for Environmental Health of "Fodor József" National Centre for Public Health, Gyali ut. 2-6, H-1097, Budapest, Hungary*

Abstract

Modelling investigation have been performed for determination of estimated concentrations of pesticides in water and for hazard categories. Predicted environmental concentration (PEC) and hazard categories for aquatic life as well as estimated human daily intake were evaluated by USES model for 275 pesticides and by HESP 2.10b software for 99 of them.
Comparison of hazard classification in line with criteria proposed by J.Linders showed that hazard categories obtained by HESP are higher than by USES. The results demonstrated that both models might be used for ranging pesticides by hazard. The appropriate use and limitations of the models should be emphasised.

1. Introduction

Classification of pesticides by hazards is not a novelty. The World Health Organisation classified solid and liquid forms of pesticides according to their acute toxicity by ingestion, inhalation and dermal absorption in 1972 [1]. Since then many different hazard classifications were developed according to special physico-chemical, toxicological, ecotoxicological properties and behaviour of pesticides in the environment as leaching, biological and non-biological degradation, bioaccumulation and undesirable side effects on fish, honey-bees, birds, wild animals and so on.
These classifications are mostly based on one specified adverse property and none of them takes into account the potential exposure to living systems.
Therefore we used environmental exposure and risk assessment models for integrated hazard assessment of pesticides.

J.B.H.J. Linders (ed.), Modelling of Environmental Chemical Exposure and Risk, 97–103.

2. Applied models

We studied about 89 pesticides used in Central and Eastern European countries. All physical, chemical and toxicological data were retrieved from Pesticide Manual [2] and the application rate of pesticides expressed in kg/hectare was taken from the Basic Technical Dossier according to Hungarian technological requirements [3]. Hazard evaluation was performed by USES [4] and HESP models [5].
The Uniform System for the Evaluation of Substances (USES) has been developed for screening and quantitative risk evaluation on both existing and new substances, with special emphasis on pesticides. The endpoint of USES (version1.0) is a quantitative comparison of the Predicted Environmental Concentration to Predicted No Effect Concentration. In our study we did not apply so called assessment factors to ecotoxicological data for extrapolation purposes (from for example an LC_{50} value to a No Effect Concentration). We simply or directly used the lowest lethal or effective concentration for the comparison to the Predicted Environmental Concentration.
We applied in our study the classification criteria proposed by Jan Linders *et al.* in 1994 [6] based on PEC/PNEC ratio as shown in table 1.

TABLE 1. Hazard for water organisms (algae, crustaceans and fish)

PEC/L/E/C_{50}	Classification
less 0.01	N- negligible
0.01-0.1	S- small
0.1-1	P- present
1-10	L- large
more/equal 10	VL- very large

The HESP (Human Exposure to Soil Pollutants) model is directed towards assessment of human exposure to chemicals, which are present in the environment as soil pollutants. The concept of the program was published by Poels *et al.* in 1991. [5]
HESP contains different modules for the evaluation of the distribution over soil fractions, for describing transfer processes and the calculation of direct and indirect human exposure. According to the default parameters HESP is a conservative model.

3. Results

Ranking of pesticides by hazard to fish. Figure 1 shows that 92 % of the investigated herbicides and fungicides belongs to the class of negligible hazard to fish evaluated by the USES model. According to the calculation by the HESP model only 69 % of these pesticides belong to the category of negligible hazard to fish with the assumptions of correct application rate.

It is evident that insecticides show higher hazard to fish and 6 to 9% of the investigated insecticides express large hazard to fish.

Comparison of the calculated total human daily dose with ADI. Comparison of averaged human daily intake calculated by USES and HESP to ADI values [7] was performed on 14 herbicides, 22 fungicides and 26 insecticides (Table 2.).

TABLE 2. Number of pesticides in different groups of human hazard

ADD/ADI	> 0.10		0.10 – 0.01		0.01 – 0.001		< 0.001	
	USES	HESP	USES	HESP	USES	HESP	USES	HESP
Herbicides n=14	4	9	2	4	4	1	4	0
Fungicides n=22	0	12	6	3	5	4	11	3
Insecticides n=26	2	12	6	7	7	4	11	3

ADD Average Daily Dose
ADI Acceptable Daily Intake

Calculation with the HESP model demonstrated that 9 herbicides exceeded the ADI. They are as follows: acetochlor, bentazone, dichlobenil, diquat, diuron, linuron, metamitron, simazine and tribenuron. (figure 2).

The ADI is exceeded by the following 5 fungicides - carbendazim, fentin-acetate, dodine, propamocarb and zineb and by the following 6 insecticides - bendiocarb, carbaryl, heptenophos, methiocarb, methomyl and pirimicarb (figures 3 and 4).

All total human daily intakes estimated by USES are below the ADI with the exception of diquat. Comparison of the results of this study with literature data [8] showed rather good correspondence (Table 3).

We can draw the conclusion that the assumed pesticides exposure to general population is on the level of acceptance. Our study also demonstrated that exposure models are useful tools for ranking pesticides by hazard.

TABLE 3. Comparison of human indirect daily intake calculated by USES, HESP and data of Camoni et al. (1991) calculated in % of ADI

Name of pesticide	USES	HESP	Camoni *et al.* (1991) /for Italy/
benomyl	0.01	19.5	2.5
carbaryl	0.41	770	0.4
chlopyrifos	0.04	3.8	0.1
chlopyrifos methyl	0.04	2.5	0.09
diazinon	3.04	75	0.2
dicofol	1.16	0.85	2.4
dimethoate	0.58	100	1.4
Fenitrothion	0.07	20	0.2
Lindane	0.61	3.38	0.3
mancozeb	0.04	--	8.1
phosalone	0.91	18	1.3
pirimifos-methyl	0.04	1.97	1.3
Thiabendazole	--	0.02	0.2
vinclozine	0.00	3	1. 7

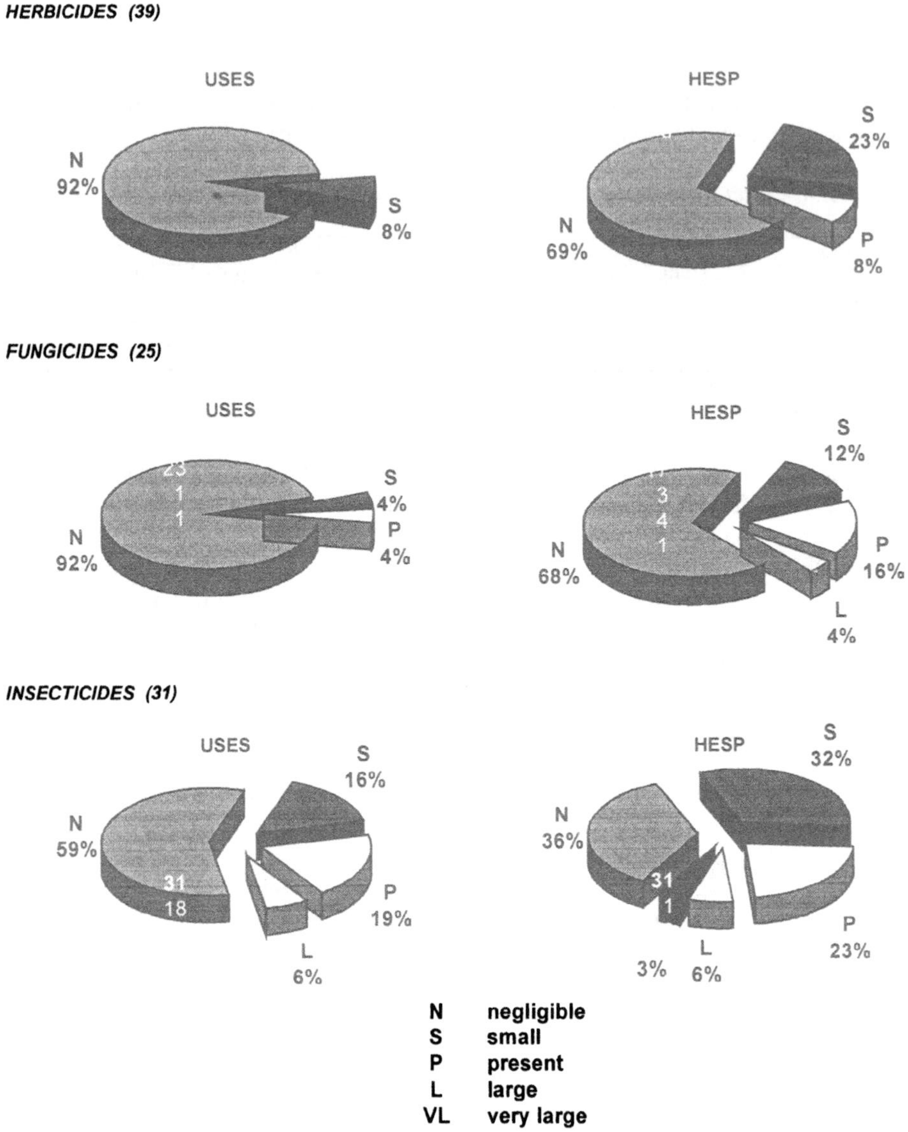

Fig. 1 Comparison of Exposure Models; PEC / lowest LC_{50} for fish

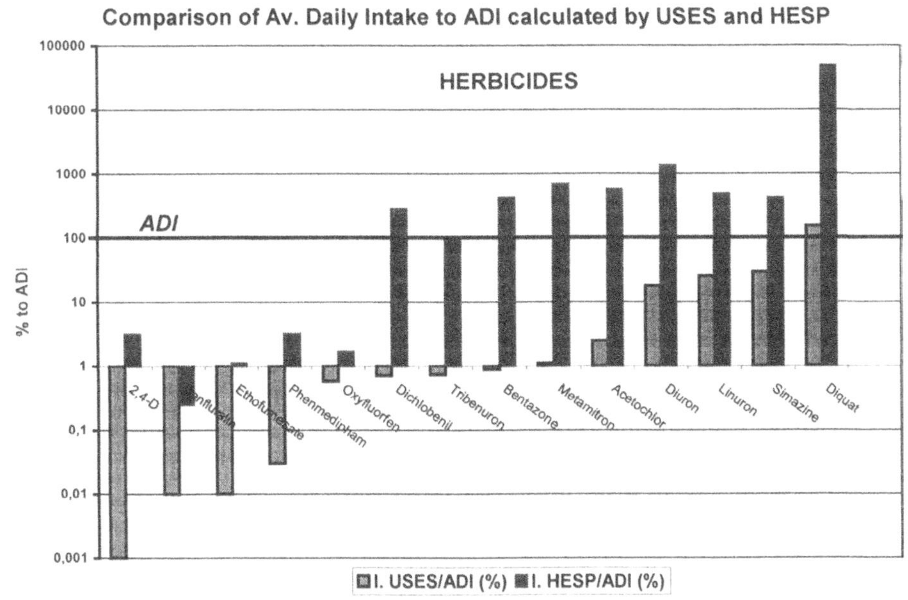

Fig. 2. *Comparison of Average Daily Intake to ADI Calculated by USES and HESP*

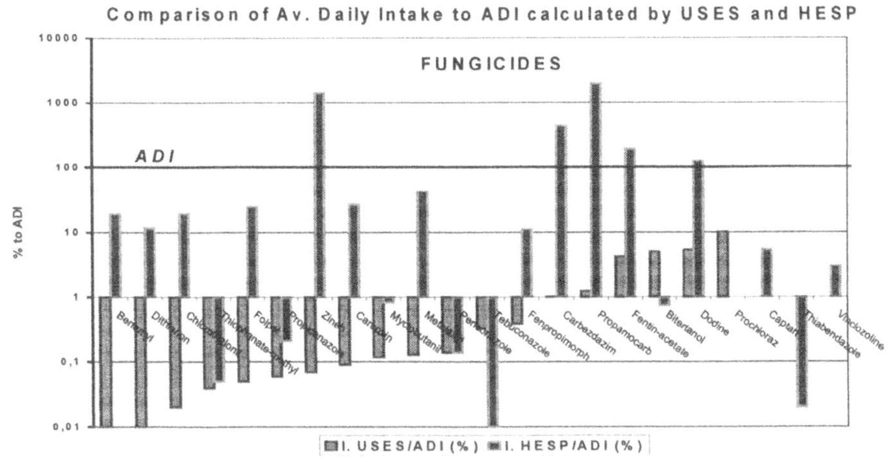

Fig. 3 *Comparison of Average Daily Intake to ADI Calculated by USES and HESP*

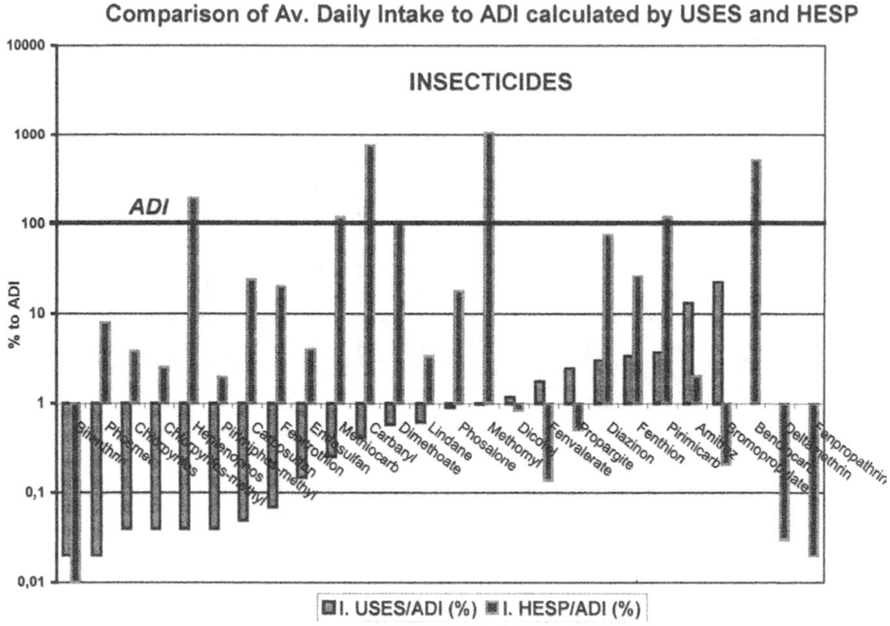

Fig.4 *Comparison of Average Daily Intake to ADI Calculated by USES and HESP*

4. References

1. WHO Recommended Classification of Pesticides by Hazard (1975) *WHO Chonicle* 29, 397-401.

2. The Pesticide Manual. (1994) 10th edition. Ed.by C.Tomlin. BCPC.

3. Engedélyezett növényvédoszerek, termésnövelo anyagok. /Permitted plant protection products in Hungary/ (1996) Ministry of Agriculture, Budapest.

4. USES-RIVM, VROM and WVC (1994). Uniform system for the evaluation of substances /USES/, version 1.0. National Institute of Public Health and Environmental Protection (RIVM), Ministry of Housing, Physical Planing and Environment (VROM), Minnistry of Welfare, Health and Cultural Affairs (WVC). The Hague, Ministry of Housing, Physical Planning and Environment, 1994. Distribution no. 11144/150.

5. Poels C.L.M., Gruntz U., Isnard P., Riley D., Spiteller M., W. ten Berge, Veerkamp W., Bontinck W.J. (1991). Hazard assessment of chemical contaminants in soil. *Chemosphere,* 23, 1, 3-24.

6. Linders J.B.H.J., Jansma J.W., Mensink, B.J.W.G., Otermann K. (1994). Pesticides: Benefaction or Pandora Box?. A synopsis of the environmental aspects of 243 pesticides. Report No 679101014. March 1994. National Institute of Public Health and Environmental Rrotection. Bilthoven. The Netherlands.

7. Lu F.C. (1995). A review of the acceptable daily intakes of pesticides assessed by WHO.*Regulatory toxicology and pharmacology* 21, 352-364.

8. Camoni et al. (1991) Cit. by Concern for Europe's Tomorrow. Health and Environment in the WHO European region. WHO European Centre for Environment and Health. Wissenschaftliche Verlagsgesellschaft mbH Stuttgart, 1995.

RECENT DEVELOPMENTS IN ENVIRONMENTAL MODELLING AT TRENT UNIVERSITY, CANADA

IAN COUSINS, MATT MACLEOD, EVA WEBSTER AND DON MACKAY
Canadian Environmental Modelling Centre, Trent University, Peterborough, Ontario, Canada K9J 7B8

Abstract

An overview of the activities at the Canadian Environmental Modelling Centre at Trent University, Ontario, Canada is presented, with emphasis on recent developments. In the last five years, user-friendly Windows-based versions of the popular multimedia fugacity models have been developed at the Centre. These models are described and applications and model validations discussed. These relatively simple, fully transparent and easily used models can be applied to rapidly evaluate environmental fate, and human and wildlife exposure. Other recent developments of the Centre reported here include: (1) the compilation of the Mackay physical-chemical property handbooks on CD ROM; (2) the inclusion of vegetation in models; (3) the development of food-web and human exposure modules to be used with the models; (4) the addition of uncertainty analysis to model outputs; (5) the use of models to define persistence and long-range transport; and (6) the development of a continent-scale model to describe chemical fate in North America.

1. Introduction

The Canadian Environmental Modelling Centre was established at Trent University in July of 1995 with the appointment of Professor Don Mackay as holder of a jointly funded NSERC (Natural Sciences and Engineering Research Council of Canada) – Chemical Industry Chair. There is active collaboration with representatives of the consortium of industrial partners, government researchers, consultants and academics.

The philosophy underlying the Centre's activities is that we must improve our ability to understand and predict the sources, fate and effects of chemical substances in the environment. Essentially, the uses and releases of chemical substances must be controlled to ensure that they pose negligible risk to the health of humans, wildlife, and the larger ecosystem of which we are only a small part. In developed countries most of

J.B.H.J. Linders (ed.), Modelling of Environmental Chemical Exposure and Risk, 105–116.

the obvious pollution sources have been regulated, but the more subtle problem remains of assessing human and wildlife exposure to the large number of organic and inorganic chemicals, which may be present at low and variable concentrations. In less developed countries, or countries in transition, pollution sources may be less well-regulated leading to higher human and wildlife exposure. Chemicals emitted in a country or region may be transported over long distances and contribute to exposure of populations living in neighbouring countries or regions.

Once emitted, contaminants can move throughout the ecosystem by a variety of pathways between atmosphere, lakes, rivers, and soils and into vegetation, animals, fish, birds, and humans. The objective of the Centre is to understand and predict quantitatively these pathways of contaminant transport and the resulting exposure, usually through the use of computer programs which mimic or model the chemicals' behaviour. When a quantitative link between sources, exposure, and risk of effects can be established it puts decision-makers in a strong position to implement measures to reduce sources to levels that avoid the problems of unacceptable contamination from excessive emissions on the one hand, and uneconomic, unnecessary regulations on the other. Such balanced regulation is best achieved through full and quantitative information about the chemicals' fate.

Models developed at the Centre are available free of charge from the Canadian Environmental Modelling Centre website, http://www.trentu.ca/envmodel, and have proven to be extremely popular, with over 6 downloads per day to more than 40 countries worldwide.

2. The fugacity concept

Most models developed at the Centre employ the fugacity concept, first introduced by G. N. Lewis in 1901 as an equilibrium criterion that was a convenient surrogate for chemical potential. The fugacity concept has proven to be an enlightening and elegant method of calculating multimedia equilibrium partitioning of organic contaminants in the environment. Pioneering work in this area was conducted by Mackay and Paterson [1,2] at the University of Toronto, Canada. Details of the fugacity approach can be found in a more recent text [3].

3. Software currently available from the Centre

Multimedia fugacity models exist in four levels of complexity with each increasing level introducing new data input requirements, and providing a more complete description of environmental fate. The models are packaged as user-friendly, Windows-based software with some representative chemicals and environmental characteristics, but users are encouraged to enter new chemicals and modify, or customise, the modelled environment.

3.1 LEVEL I

Level I fugacity calculations describe a situation in which a fixed quantity of chemical is allowed to come to thermodynamic equilibrium within a closed and defined environment. The Level I software program guides the user sequentially through the selection of chemical and environmental properties then computes chemical partitioning. There is no resistance to chemical migration between environmental phases, and each phase is considered to be homogeneous and well mixed. This calculation is intended to provide a general impression of the environmental media into which a chemical is likely to partition based principally on physical-chemical properties such as vapour pressure, octanol-water partition coefficient and water solubility. The concept of equilibrium bioconcentration as controlled by the octanol-water partition coefficient is used. A representative output diagram for hexachlorobenzene from the Level I simulation is shown in Figure 1. Details of the simulation can be viewed in tables in the program or saved in a file.

3.2 LEVEL II

Level II calculations describe a situation in which the chemical is continuously discharged at a constant rate and reaches steady state and equilibrium conditions at which the emissions are balanced by advective and degradation losses. The Level II software is consistent in appearance and operation with Level I. Degradation half-lives, not required for Level I, must be supplied. New information obtained includes the overall environmental persistence of a chemical, and the relative importance of chemical degradation versus transfer to adjacent regions as mechanisms of removal.

3.3 LEVEL III

Level III fugacity calculations introduce resistance to chemical migration between environmental compartments, and are the most useful and enlightening evaluative calculations. Level III describes a situation in which the chemical is continuously discharged at a constant rate to one or more of the four environmental compartments and achieves a steady state, non-equilibrium condition. Additional input parameters that characterise the environment are required to calculate transfer rates between environmental media. The Level III calculation shows the dependence of overall environmental fate on media of release and highlights the dominant intermedia transport pathways (e.g. air-water or air-soil exchange).

Level IV unsteady-state (or dynamic) models can also be formulated to analyse short-term (e.g. seasonal) effects on environmental concentrations, or to determine the effect of changing environmental emissions over a period of decades. There is no software available on the Centre's website for carrying out Level IV calculations; it is under development.

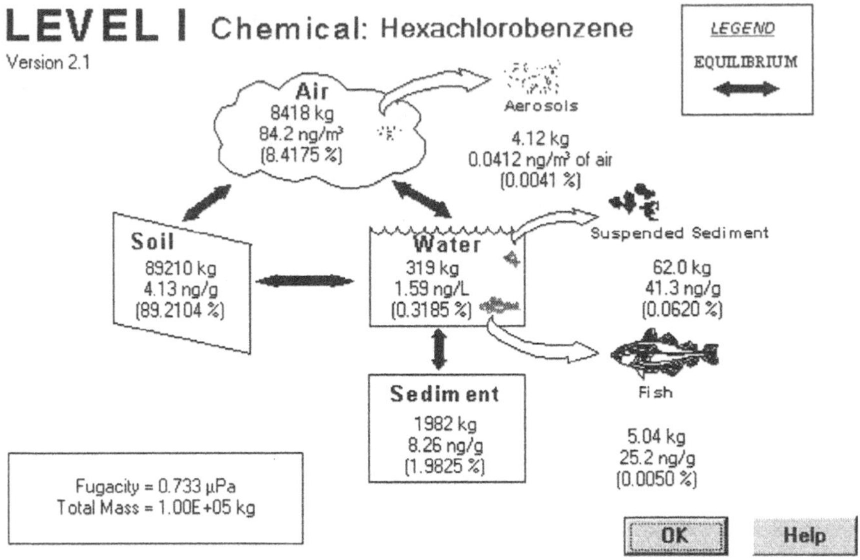

FIGURE 1. Screen capture of Level I output diagram for hexchlorobenzene

3.4 EQC

The Equilibrium Criterion (EQC) evaluative model deduces the fate of a specified chemical in an evaluative environment of fixed properties [4]. The aim is to provide a "benchmark" environmental fate profile for a chemical that can be compared to the fates of other chemicals in the same model world. The model treats chemicals that fall into three categories; (1) chemicals that partition into all environmental media, (2) those that are involatile, and (3) chemicals that are insoluble in water. Level I, II and III assessments can be made simultaneously.

3.5 CHEMCAN

ChemCAN is a steady-state Level III model that describes the multimedia behaviour and fate of organic and inorganic chemicals in 24 defined regions of Canada [5]. It is easily adapted to other regions and has recently been applied to part of Canada [6], France [7] and Germany [8]. In addition to the media in Level III and EQC, ChemCAN treats vegetation, groundwater and coastal water. An output diagram for a simulation in which 1000 kg/year of hexachlorobenzene is emitted to the air, water and soil compartments of the "Ontario-mixed wood plain" region is shown in Figure 2.

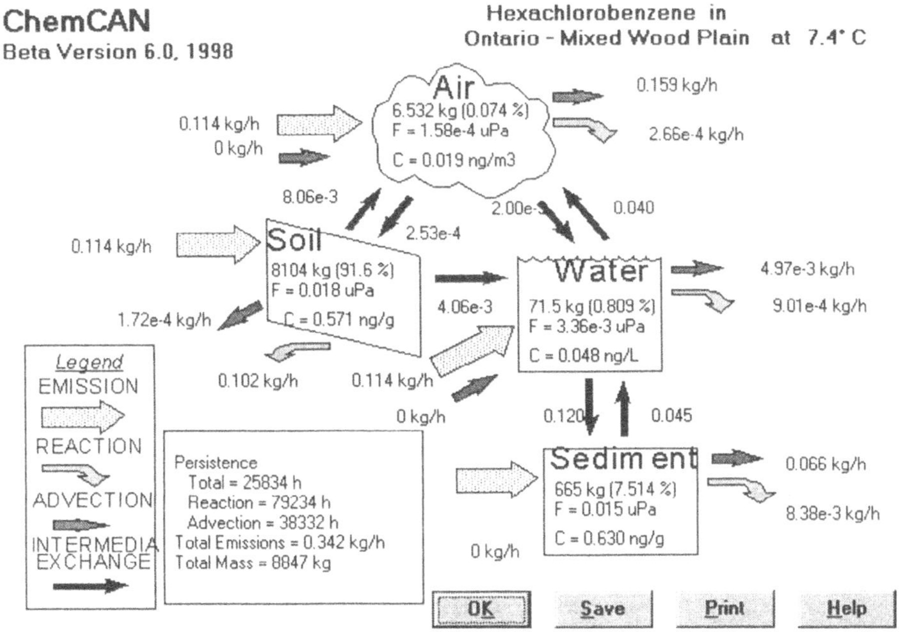

FIGURE 2. Screen capture from a typical ChemCAN simulation

3.6 QWASI

The "Quantitative Water Air Sediment Interaction" or "QWASI" model was designed to assist in understanding organic chemical fate in lakes [9], and a DOS version of the model is currently available from the Centre's website. More recently QWASI has been adapted to describe the fate of organics in river systems [10] and to describe the fate of speciating chemicals [11,12]. QWASI describes the steady-state behaviour of an organic chemical in an aquatic system subject to chemical inputs by direct discharge, inflow in rivers/tributaries, and deposition from the atmosphere. Chemical is removed from the water by evaporation, irreversible reaction in the water and sediment, outflow in the water, and sediment burial. The mass balance equations for the well-mixed water column and the well-mixed layer of surficial sediment include sediment-water exchange by diffusion, deposition, and resuspension. Three compartments are treated; air, water and sediment, but the water compartment can be subdivided to account for thermal stratification. Rivers are modelled as connected QWASI lake models with transport between adjacent water compartments by advective and diffusional flow. The degree of segmentation is set by the modeller and must account for the hydrodynamics of the river system and the location of measured data used for validation. A Windows-based version of QWASI is under development.

4. Model validation

When the model environment is purely evaluative or hypothetical there is no possibility of validation, but the equations used to describe partitioning, transport, and transformation are identical to those that have been used successfully in models of chemical fate in more defined environments such as ponds or soil plots. Models that describe the fate of chemicals in a particular environment or region (e.g. ChemCAN) can only be partially validated by comparison with available data on ambient environmental concentrations. It is expected that the agreement of model predictions with environmental concentration data will be approximate, i.e. a factor of 2 or 3 rather than, for example, plus or minus 20%. This level of accuracy, however, is often adequate to show whether concentration and hence an exposure is consequential or not. In risk assessment applications, toxicity data are rarely more accurate. Further, environmental concentrations can readily vary by such factors spatially and temporally.

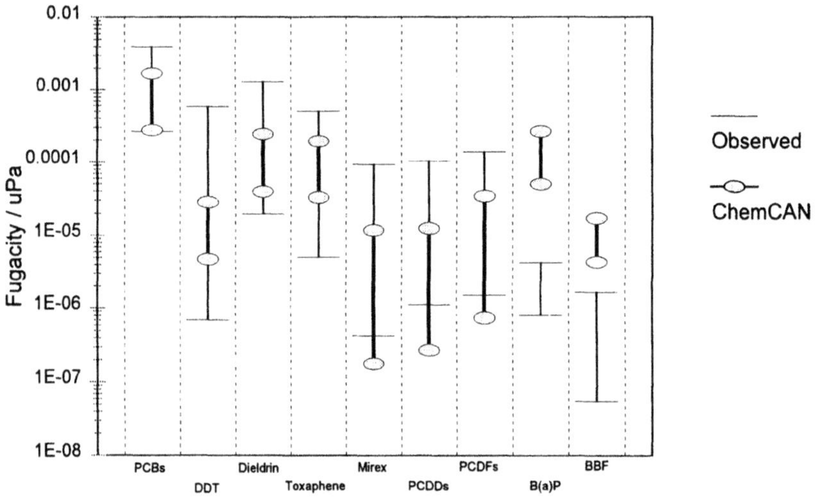

FIGURE 3. Plot of observed and calculated levels of priority pollutants in southern Ontario air

Recently, a partial validation of the ChemCAN model was attempted for benzene, the chlorobenzenes and 13 other organic pollutants in the Southern Ontario environment [6]. In general, satisfactory agreement within a factor of 2-3 was found between predicted and observed environmental concentrations (see Figure 3). Potential causes for over-estimation of concentrations of chemicals such as B[a]P include the absence of metabolism and seasonal effects in the model.

Further validation of multimedia environmental models is needed. Therefore, groups measuring environmental concentrations need to collaborate more with model developers. In addition, modellers around the world should engage in inter-model comparison exercises to ensure that the large number of models now available are providing similar and correct predictions.

5. Model applications

A successful model can be used to derive likely environmental concentrations and exposures for assessments of new chemicals and to focus funding of monitoring projects of existing chemicals of commerce. For the environmental fate assessment of chemicals Mackay *et al.* [13] have suggested a five-stage strategy that incorporates the use of multimedia models. The five stages are:

(1) Chemical classification: chemicals are selected and physical-chemical properties characterised.

(2) Acquisition of discharge data and environmental concentrations: obtain data on chemical production, emissions, and background concentrations.

(3) Evaluative assessment of chemical fate: deduce the general features of chemical behaviour in a generic environment at 25°C. The EQC model is recommended for this stage of the assessment.

(4) Regional or far-field evaluation: estimate chemical fate in region of 10^4 to 10^6 km^2 and identify local situations deserving further study. The ChemCAN model is an example of a regional model that can be used in this stage of the assessment.

(5) Local or near-field evaluation: evaluate local behaviour in regions and media of high use and exposure. For an aquatic, local assessment, QWASI is recommended.

For risk assessments a sixth stage is needed to determine the environmental exposure and probability of associated effects to humans and/or wildlife. MacLeod and Mackay [6] demonstrated that a multiple pathway human exposure model can be readily interfaced to a regional model such as ChemCAN. The output from this model is a total predicted daily intake (PDI), which can be compared to an allowable daily intake (ADI), formulated by regulators using toxicity data.

Researchers find our available software useful for identification of key environmental processes that need to be quantified more accurately. Sensitivity tests can be used to determine which input parameters are most, or least, important. Overall environmental persistence can be estimated and the key loss processes affecting a chemical's persistence can be determined. Models are also useful environmental management tools for regulators and industry decision-makers. They can be used to link loadings to concentrations quantitatively, and therefore identify key sources. A Level IV unsteady-state model can be used to assess the effect of changing emission patterns over time and tentatively estimate environmental concentrations in the future.

6. Recent developments at the Centre

The Centre is continually working to improve existing models and develop novel modelling approaches. A selection of notable current efforts and developments are discussed below.

6.1 PHYSICAL-CHEMICAL PROPERTIES

The 5-volume "*Illustrated Handbooks of Physical-Chemical Properties*" has recently been successfully converted to CD ROM format and is now available from CRC Press [14]. This CD allows modellers easy access to the chemical input data needed to run model simulations of many organic compounds.

6.2 INCLUSION OF VEGETATION

Multimedia models that describe the environmental fate of organic contaminants do not generally include vegetation compartments. There are several reasons for this omission:

(1) For many organic contaminants the quantities partitioning to vegetation are believed to be small compared to the amounts in other media and thus vegetation may not significantly influence environmental fate.

(2) There is a lack of experimental measurements on uptake of organic contaminants by vegetation, thus the uptake kinetics and equilibria are poorly understood making it difficult to justify model equations.

(3) It is not obvious how roots, stems, foliage and fruit should be combined, or segmented and it is uncertain how different types of vegetation (e.g. trees and grasses) should be treated or combined into classes.

In spite of these difficulties, multimedia models are justifiably criticised for not including vegetation, which is clearly an important aspect of the terrestrial surface; approximately 80% of the Earth's land surface is covered by vegetation [15]. It has been suggested that vegetation, especially forests, could be an important sink for organic compounds [16], as a result of the high reactivity of some compounds on vegetation surfaces and the array of available deposition surfaces in vegetation canopies.

A multimedia fugacity model has recently been developed [17] that includes vegetation as a primary medium in the mass balance calculations. In this model vegetation is divided into four broad classes (deciduous trees, coniferous trees, grasses and agricultural crops) and each vegetation compartment is paired to a corresponding soil compartment. The new parameters to incorporate vegetation into the multimedia model have been formulated, although there are still uncertainties about some of the equations used. It is hoped that new experimental data from our collaborators (principally Dr. Kevin Jones at Lancaster University in the UK) will help us to better understand plant uptake processes and improve the current model.

6.3 ECOWEB

A wildlife food-web model (EcoWeb) is currently being developed that can be used to evaluate the potential for bioaccumulation, biomagnification and food-web transfer in combined or individual aquatic and terrestrial systems. The model will accept as inputs the concentration estimates from the regional models, or monitoring data, and from them deduce chemical partitioning into a range of biota including invertebrates, fish, mammals, birds, and vegetation. The model will include the potential for metabolic degradation in all biota. The preliminary version that has already been developed [18] provides assurance of feasibility. However, the model still needs to be applied to several food webs that are well understood and have environmental concentrations available for all biota. This validation work is currently being undertaken using datasets collected from several of the Centre's collaborators around the world.

6.4 UNCERTAINTY ANALYSIS

Efforts are underway to develop methodologies for analysis of uncertainty in output from mass balance models and to quantitatively identify the input parameters and fate process descriptions that are principal sources of uncertainty. Monte Carlo analysis is a widely recognised method of assessing uncertainty in mathematical models, and has been applied to a version of the QWASI model by Thompson *et al.* [19]. For complex regional contaminant fate models such as ChemCAN, however, Monte Carlo analysis is computationally intensive and significantly increases computing time. In addition, the results of the analysis can be difficult to interpret by users unfamiliar with statistical mathematics. A deterministic approach to uncertainty analysis of contaminant fate models is currently under development at the Centre. The approach provides easily interpretable estimates of uncertainty that are consistent with Monte Carlo analysis, without significant increases in model complexity and computation time. This screening level uncertainty analysis identifies important sources of model uncertainty, and can be used by researchers to guide more complex Monte Carlo analysis where required.

6.5 PERSISTENCE AND LONG-RANGE TRANSPORT

The role of multimedia mass balance models in assessments of environmental persistence is being investigated. Initial recommendations focussed on the use of Level III fugacity calculations, which include the effects of chemical partitioning and mode-of-entry information [20]. A method for rapid screening of the thousands of chemicals to be assessed by government regulators has recently been developed [21]. This rapid screening uses Level II calculations and an examination of the chemical's properties to determine the relative importance of individual media half-lives. It is hoped that this screening approach will be used to prioritise chemicals for the more detailed Level III assessment.

A new model to assess long-range transport potential of chemicals has been developed [22]. It calculates the "characteristic travel distance" for chemicals in air or water based on partitioning characteristics and an average flow velocity for air and water. Chemicals identified by the model as having long characteristic travel distances include HCB, HCH and toxaphene, in good agreement with known globally distributed contaminants. A sample of the results is shown below in Table 1 with no units since values are for relative ranking only.

TABLE 1. Persistence and long-range transport distances for a range of priority organic pollutants

Chemical	Persistence	Travel distance
HCB	2300	1300
α-HCH	220	130
chlorobenzene	15	51
Hexa-PCB	4700	32
Toxaphene	2100	29
Biphenyl	5.1	16
Dieldrin	240	15
pp'-DDT	1500	7.7
Benzene	1.5	5.1
OCDD	4600	4.7

7. The Future

A major initiative is to develop a multi-segment regional model of chemical fate in the whole of North America (i.e. including the USA and Mexico) with region-to-region transport. Tentatively this model will contain 15 regions in Canada, 12 in the USA, 4 in Mexico and 8 surrounding ocean regions. Regional segmentation will be based on ecosystem characteristics and population distribution. Each segment will have defined climatic, soil, land-use, and hydrological characteristics, and segment-to-segment transport in air and water will be quantified. The net result will be that an emission or source in any one segment can be tracked to give quantities and concentrations transported to all other segments. It will thus directly assess trans-boundary pollution and long-range transport.

8. Acknowledgements

We are grateful to NSERC and the consortium of chemical companies that provide finances to support the Canadian Environmental Modelling Centre and to the NATO Science Program for inviting us to participate in this workshop and especially to Fina Kaloianova for assistance.

9. References

1. Mackay, D. and Paterson, S. (1981) Calculating fugacity, *Environ. Sci. Technol.*, 15, 1006-1014.

2. Mackay, D. and Paterson, S. (1982) Fugacity revisited, *Environ. Sci. Technol.*, 16, 654A-660A.

3. Mackay D. (1991) Multimedia environmental models: The fugacity approach, CRC Press LLC, Lewis Publishers, Boca Raton, FL, USA.

4. Mackay, D. Di Guardo, A., Paterson, S. and Cowan, C. (1996) Evaluating the environmental fate of a variety of types of chemicals using the EQC model, *Environ. Toxicol. Chem.*, 15, 1627-1636.

5. Mackay, D., Webster, E., Di Guardo, A., Paterson, S., Kane, D. Assessing chemical fate in regional environments: The ChemCAN model, *Environ. Sci. Technol.* (submitted).

6. MacLeod, M. and Mackay, D. (1999) An assessment of the environmental fate and exposure of benzene and the chlorobenzenes in Canada, *Chemosphere*, 38(8), 1777-1796.

7. Devillers, J. and Bintein, S. (1995) CHEMFRANCE: A regional Level III model fugacity model applied to France, *Chemosphere*, 30, 457-476.

8. Kühne, R., Breitkopf, C. and Schüürmann, G. (1997) Error propagation in fugacity Level III models in the case of uncertain physicochemical compound properties,*Environ. Toxicol. Chem.*, 16, 2067-2069.

9. Mackay, D., Joy, M. and Paterson, S. (1983) A quantitative water, air, sediment interaction (QWASI) fugacity model for describing the fate of chemicals in lakes, *Chemosphere*, 12: 981-997.

10. Mackay, D., MacLeod, M., Milford, L., Webster, E., Bentzen, E., Hickie, B., Sharpe, S., Thompson, S., Lee, K. and Hellou, J. (1998) Towards an ecosystem-wide model of contaminant fate in the Great Lakes - St. Lawrence Regions, *Great Lakes Res. Rev.*, 4, 9-14.

11. Diamond, M., Mackay, D. and Welbourne, P.M. (1996) Models of multi-media partitioning of multi-species chemicals: The fugacity/aquivalence approach, *Chemosphere*, 25, 1907-1921.

12. Woodfine, D., Seth, R., Havas, M. and Mackay, D. Simulating the Remediation of Metal Contaminated Lakes using a Modified QWASI Model, *Chemosphere* (in press).

13. Mackay, D. Di Guardo, A., Paterson, S., Kicsi, G. and Cowan, C. (1996) Assessing the fate of new and existing chemicals: a five-stage process, *Environ. Toxicol. Chem.*, 15, 1618-1626.

14. Mackay, D., Shiu, W-Y. and Ma, K-C. (1999) Physical-Chemical Properties and Environmental Fate Handbook, CRC netBASE CD-ROM, Chapman and Hall/CRC Press, Boca Raton, FL, USA.

15. Simonich, S.L. and Hites, R.A. (1994) Vegetation-atmosphere partitioning of polycyclic aromatic hydrocarbons, *Environ. Sci. Technol.*, 28, 939-943.

16. McLachlan, M.S. and Horstmann, M. (1998) Forests as filters of airborne pollutants: a model, *Environ. Sci. Technol.*, 32, 413-420.

17. Cousins, I.T. and Mackay, D. Including vegetation in multimedia fugacity models: when are they needed and how can it be done? *Chemosphere* (submitted).

18. Sharpe, S. and Mackay, D. A framework for evaluating bioaccumulation in food-webs,*Environ. Sci. Technol.* (in press).

19. Thompson, S., Mackay, D. and MacLeod, M. A modelling strategy for planning the virtual elimination of persistent toxic chemicals from the Great Lakes: an illustration of four contaminants in Lake Ontario, *J. Great Lakes Res.* (in press).

20. Webster, E. Mackay, D. and Wania, F. (1998) Evaluating environmental persistence, *Environ. Toxicol. Chem.*, 17, 2148-2158.

21. Gouin, T., Mackay, D., Webster, E. and Wania, F. Screening chemicals for persistence in the environment,*Environ. Sci. Technol.* (published on the web January 2000).

22. Beyer, A., Mackay, D., Matthies, M., Wania, F. and Webster, E. Assessing long-range transport potential of persistent organic pollutants, *Environ. Sci. Technol.* (published on the web January 2000).

MACRO: A PREFERENTIAL FLOW MODEL TO SIMULATE PESTICIDE LEACHING AND MOVEMENT TO DRAINS

SABINE BEULKE[a], COLIN D. BROWN[a], NICHOLAS J. JARVIS[b]
[a]*Soil Survey and Land Research Centre, Cranfield University, Silsoe, Bedford, MK45 4DT, UK;*
[b]*Department of Soil Sciences, Swedish University of Agricultural Sciences, PO Box 7072, 750 07 Uppsala, Sweden*

Abstract

Preferential flow is an important process for pesticide transport through a wide range of soils and this phenomenon should be considered in regulatory assessments for pesticide leaching to ground and surface waters. One of the simulation models available for these purposes (MACRO) and its database version, which allows automatic parameter selection (MACRO_DB), are described in this paper. The main concepts of the models and their current role within pesticide regulation are reviewed. A study on the evaluation of MACRO and MACRO_DB against four UK datasets is summarised. This suggested that MACRO is likely to continue to be the preferred preferential flow model for regulatory purposes. The model is user-friendly and well-documented and was shown to predict leaching through a wide range of soils with some degree of accuracy. However, parameter selection for MACRO requires expert soil physics knowledge and the model should only be applied by an experienced user. A comprehensive calibration step should be included wherever possible. The philosophy behind MACRO_DB to automatically select parameters from basic soil information is commendable, but some of the estimation algorithms need to be improved. Overall, the importance of preferential flow modelling within pesticide registration is likely to increase as more results of validation studies and guidance to the user become available.

1. Introduction

Pesticide leaching models are widely used within European registration to assess potential losses of pesticides to ground and surface waters. Generally, the aim is to carry out realistic worst-case assessments. This involves the selection of soil scenarios, which give a worst-case situation relevant to the proposed conditions of use. In Germany, leaching to groundwater through a sand or sandy loam soil (< 30% clay + silt and < 10% clay) with an organic carbon content of less than 1.5% is assessed [1]. A

J.B.H.J. Linders (ed.), Modelling of Environmental Chemical Exposure and Risk, 117–132.

similar soil is used within the Dutch standard scenario for modelling [2]. These scenarios are likely to represent a worst-case for pesticide transport via chromatographic flow through disturbed soils due to their relatively large hydraulic conductivity and limited sorption and degradation capacity. However, experimental studies suggest that these soil types may not give a worst-case situation for leaching through undisturbed soils. Brown *et al.* [3] compared the leaching of pesticides including isoproturon through lysimeters from five contrasting soil types. A structureless sandy soil, which meets German guidelines for a worst-case soil for leaching was found to be less vulnerable to pesticide leaching than a structured clay loam and a weakly structured deep loam. This was attributed to preferential flow through the structured soils.

Preferential flow refers to a situation where water infiltrating a soil does not equilibrate with the resident soil water, but flows rapidly to depth through a small portion of the soil volume. For example, in structured soils, shrinkage cracks, earthworm channels or root holes may operate as flow pathways in which water moves rapidly downwards and by-passes the denser soil matrix [4]. Thereby, chemicals dissolved in the infiltrating water may quickly move out of the biologically and chemically reactive topsoil into the subsoil where sorption and degradation is limited. The risk of leaching of those chemicals to ground and surface waters is thus enhanced. Preferential flow in the form of unstable wetting fronts or "fingering" may also occur in unstructured sandy soils due to local density and textural differences.

There is evidence to suggest that preferential flow may be an important process for pesticide transport through a wide range of soils. Rapid movement through heavy clay soils to sub-surface drains appears to be a dominant pathway for pesticide transport to surface waters [5], [6], [7]. Pesticides by-passing the matrix of intermediate-textured soils overlying aquifers may cause transient large concentrations in groundwater [8], [9]. Flury *et al.* [10] identified preferential flow processes in thirteen of fourteen Swiss arable soils studied and concluded that the occurrence of preferential flow is the rule rather than the exception. There is thus a need to include preferential flow into regulatory risk assessments for pesticides and a demand for models, which explicitly account for this phenomenon. A number of preferential flow models have been developed including CRACK-NP [11], PLM [12], MACRO [13], RZWQM [14], and SWAT [15]. This paper focuses on MACRO, because it is the preferential flow model, which is most widely used for regulatory purposes. It reviews the concepts built into the model, its documentation and ease of parameterisation. The role of the MACRO model within European pesticide registration is also discussed. In addition to MACRO, the database version of the model, which allows automatic selection of soil hydraulic and crop parameters (MACRO_DB), is described.

2. Description of the MACRO model

MACRO [13] is a physically-based preferential flow model. In contrast to conventional models which do not simulate preferential flow, MACRO divides the total soil porosity into a slow flow domain (the soil matrix or micropores) and a domain of faster flow (the macropores) as shown in Figure 1.

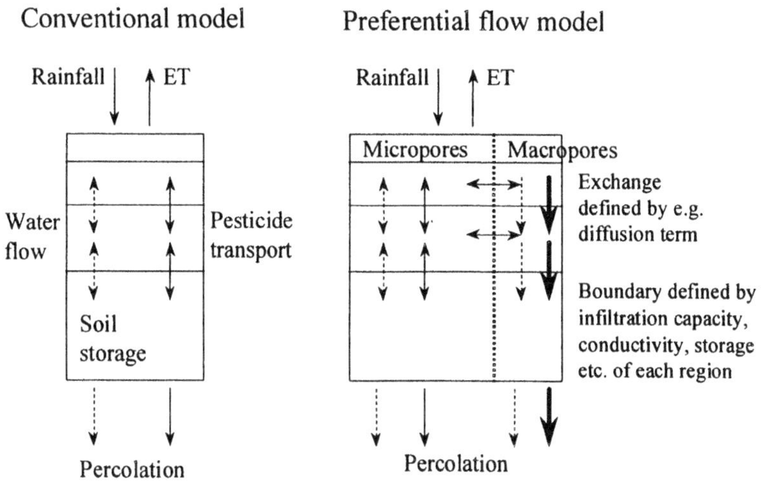

Figure 1 *Basic structure of a conventional pesticide leaching model and the preferential flow model MACRO*

The main features of Version 4.1 of the MACRO model are summarised below (partly adapted from Boesten *et al.* [16] and Adriaanse *et al.* [17]).

Soil model

Soil column divided into homogeneous layers of variable thickness

Hydrology model

Two-domain model with total pore space divided into macropores and micropores; solution of Richards' equation within micropores, capacitance approach within macropores

Preferential flow

Division between domains defined by boundary water tension, water content and hydraulic conductivity. Preferential flow initiated when boundary water content is exceeded

Exchange of water between domains

Water enters macropores when infiltration capacity or boundary water content of the micropores is exceeded. Flow in reverse direction according to first-order process using an effective aggregate half-width.

Potential and actual evapotranspiration

Input of potential evapotranspiration (PET) data or estimation using Penman-Monteith's equation; PET partitioned between leaf and crop in proportion to leaf area index. Maximum possible flux of water considered as well as aeration and drought stress.

Overland flow

Considered, but only as a means to remove excess water. Not recommended for predictive use

Drainage

Seepage potential theory

Solute transport

Convection-dispersion equation in micropores, mass flow in macropores

Exchange of solute between domains

Diffusion and mass flow, with aggregate half-width as an important parameter

Pesticide sorption

Freundlich; sorption sites partitioned between micro- and macropores; different Kf for each layer, but single Freundlich exponent; Kf can be reset to a new value at any time during the run to account for time-dependent sorption

Pesticide degradation

First-order; temperature and moisture effects may be modelled; different rate constants for solid and liquid phases of micro- and macropore domain possible; different rate constants for each layer; fate of metabolites may be simulated

Plant shoot growth

Leaf area indices and form factors specifying growth curve; starting date of regrowth of winter-sown crops in spring may be specified

Plant root growth

Linear interpolation between minimum at emergence and maximum when the crop has its maximum leaf area; root volume distributed logarithmically with depth

The boundary between the macropore and micropore domain is defined within MACRO by the boundary water content, water tension and hydraulic conductivity. Water enters the macropores as soon as the infiltration capacity of the micropores (upper soil layer) or the boundary water content (remaining profile) are exceeded. By varying the input parameters, the MACRO model can be set up to simulate a soil dominated by preferential flow, a soil with no preferential flow at all or any combination of flow types between these two extremes. The main application of MACRO is to simulate flow through structured clay soils where rapid movement of water and solutes through cracks and fissures is important. The model can also be

applied to intermediate loam soils where earthworm and root channels may deliver water and solutes rapidly to depth. The processes of finger flow and funnel flow in coarse-textured soils cannot be simulated with MACRO.
MACRO and relevant documentation can be downloaded from the website www.mv.slu.se/bgf/Macrohtm/download.htm (Swedish University of Agricultural Sciences, Uppsala). MACRO is well documented and has a user-friendly front-end. Some guidance on selection of model parameters and reasonable ranges is given. However, robust parameter selection for MACRO is rather difficult and output is particularly sensitive to changes in some of the more problematic parameters (*e.g.* aggregate half-width, position of the boundary between micropore and macropore domains, proportion of sorption sites within each). Considerable previous experience with MACRO is required and a comprehensive calibration step should be included wherever possible.

3. Description of MACRO_DB

The philosophy behind MACRO_DB [18] is to reduce the problems of robust parameter selection which have been identified for MACRO. It has been designed for management applications by the non-specialist user in making exposure and risk assessments for pesticides. The structure of MACRO_DB is schematically presented in Figure 2.

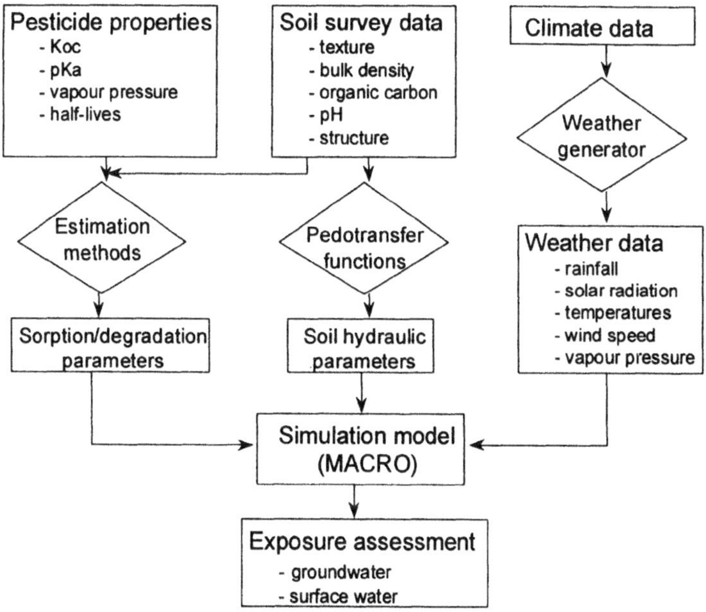

Figure 2. Structure of MACRO_DB

MACRO_DB links various data sources to the MACRO model by the use of parameter estimation algorithms. The databases provided include pesticide properties, soils, cropping and weather. Currently, the system allows access to the PETE database [19], which contains information for over 600 common pesticides. The pesticide properties required for model input and stored in the database are: ionic state (non-ionic, acid, base or cation), octanol-water partition coefficient (or Koc for non-ionic compounds), pKa (for acids and bases only), vapour pressure at a given temperature, topsoil half-life at a given temperature and optimum water content, degree of delocalization (for bases and cations only) and valency (for cations, mono- or di-cations). The soils databases which can be accessed by MACRO_DB are SEISMIC [20] containing ~400 UK soils and MARKDATA (26 Swedish soils). In addition, information for three German soils, two Swedish soils and a French soil are stored in a database that the user can develop further. The soils data stored are: horizon designation and thickness; and for each horizon in the profile texture class (either U.S.D.A. or U.K. system) or particle size distribution, organic carbon content, bulk density, pH and the soil structure description (according to FAO system). The cropping scenarios currently provided by MACRO_DB are:

Crop	**Location**
Winter wheat	Karlsruhe, Germany
Winter wheat	Neuenkirchen, Germany
Winter wheat	Rothamsted, U.K.
Winter wheat	Silsoe, U.K.
Winter wheat	Skåne, Sweden
Winter wheat	Vredepeel, Netherlands
Winter barley	Watnall, U.K.
Sunflower	Karlsruhe, Germany
Sugar beet	Neuenkirchen, Germany
Spring rape	Birmingham, U.K.
Spring oats	Uppsala, Sweden
Spring barley	Uppsala, Sweden
Potato	Rothamsted, U.K.
Permanent grass	Silsoe, U.K.
Main crop potatoes	Southern Sweden
Grass	Rothamsted, U.K.
Field beans	Silsoe, U.K.
Cotton	Marismas, Spain

MACRO_DB also provides a number of weather scenarios containing information on rainfall, solar radiation, wind speed, vapour pressure or relative humidity and air temperature on a daily basis:

Braunschweig, Germany	5 years real data
Halmstad, Sweden	10 years real data
Lanna, Sweden	8 years real data
Rothamsted, UK	2 years real data
Uppsala, Sweden	26 years real data

When daily data are not available, these can be generated by MACRO_DB using an in-built weather generator from long-term average monthly climate data. The minimum requirements are: monthly rainfall and average number of raindays, mean temperature, mean wind speed, latitude and altitude.

The databases are currently being updated to include the scenarios for leaching to groundwater and surface water developed by the FOCUS work groups. In addition to using the existing pesticide, soil, cropping and weather scenarios, users can develop their own databases.

Following the selection of scenarios, MACRO_DB will automatically derive input parameters for MACRO using a combination of simple rules and pedo-transfer functions. Thus, the sensitive parameters setting the boundary between micropores and macropores and governing the rate of exchange of water and solute between regions can be set independent of any user subjectivity.

4. Evaluation of the regulatory use of MACRO and MACRO_DB

MACRO is currently the preferred preferential flow model for regulatory use and was one of the models recommended by the European FOCUS work-groups [16], [17] to predict pesticide leaching to ground and surface waters within the registration procedure. However, the regulatory use of preferential flow models or the weight which modelling results have been afforded have been restricted by a lack of information on the relative strengths and weaknesses of current models and problems of parameterisation. These problems have been emphasised by the FOCUS Leaching Models work group [16] who stated that: "Current models that consider macropore flow require that soil parameters be obtained by calibration. More advances are needed before predictions of macropore flow can be made using soil parameters in existing databases". The Surface Water Models work group [17] expressed similar sentiments: "Outputs from the macropore flow models MACRO and CRACK-NP are sensitive to parameters related to the macropore region, which are, in turn, difficult to estimate. This may lead to high levels of predictive uncertainty compared to the use of models in non-structured sandy soils." At present, the MACRO model is mainly used for higher tier modelling - for example to extrapolate results from field studies to a wider range of climatic conditions. Nonetheless, the use of MACRO or MACRO_DB to predict environmental concentrations in groundwater is requested by some regulatory authorities (*e.g.* Denmark). More simplistic models which do not simulate preferential flow (*e.g.* PELMO, PESTLA, PRZM) are used for registration in other European countries (*e.g.* Germany, The Netherlands).

Standard modelling scenarios to assess the potential for a pesticide to leach to ground and surface water are currently being developed by the FOCUS work groups. Only one of the nine proposed groundwater leaching scenarios (a silty clay loam soil from Châteaudun in France) requires the use of MACRO to account for the importance of preferential flow [21]. The remaining scenarios may be simulated with either PESTLA, PRZM or PELMO. FOCUS scenarios are also being developed for higher tier modelling of pesticide losses to surface water via drainflow. These will also be simulated with MACRO (J.M. Hollis, personal communication).

5. A case study on the regulatory use of MACRO under UK conditions

The use of MACRO (version 4.0) under UK conditions has been evaluated in a recent project by Beulke *et al.* [22] and implications for its regulatory use were discussed. MACRO and three additional preferential flow models (CRACK-NP, PLM, and SWAT) were evaluated against four datasets. Results were compared with those from a non-preferential flow benchmark (LEACHP, [23]). MACRO was evaluated both as the stand-alone model (MACRO) and as the database version MACRO_DB.

5.1 DATASETS

Data have been collected by various organisations from three field sites and one lysimeter experiment in the UK. Soils at the field sites are either clays (Brimstone Farm and Wytham) or a clay loam (Cockle Park) in which preferential flow is likely to be an important pathway for water and solute movement through the profile. Data on drainflow and associated concentrations of isoproturon from Brimstone Farm were available for four experimental plots for the first two key rainfall events of each of three successive seasons. At Wytham, isoproturon was applied in spring 1994 to a winter barley crop on a mole-drained field. Isoproturon concentrations in drainflow were recorded during key events together with the respective flow rates. Data from Cockle Park were for losses of water and pesticides including isoproturon in drainflow from a mole-drained plot between November 1990 and March 1991. The lysimeter experiment (SSLRC) was conducted with five contrasting soil types with varying texture and potential for preferential flow. Bromide and pesticides including isoproturon were applied in two successive seasons and their concentrations in leachate were monitored throughout the leaching periods. All studies were supplemented with either experimental or literature data on pesticide sorption and degradation.

5.2 MODEL EVALUATION

The models were run uncalibrated. Wherever possible, measured data were used to select input parameters for the stand-alone version of MACRO or default values were retained to avoid introducing unnecessary user-subjectivity. MACRO parameters, which are difficult to select, are those describing the soil hydraulic properties. These were derived from the measured water release curve, by expert judgement or using estimation algorithms.

Simulations with MACRO_DB retained the weather, pesticide, application and site hydrology (drain depth, drain spacing, depth of profile) parameters from the simulations with the stand-alone version of MACRO. A set of crop input parameters was selected from the validated database provided with MACRO_DB based on the closest to the desired crop and locality. Only the dates of emergence and harvest were altered to match those at each site. Soil hydraulic parameters were calculated within MACRO_DB using the automatic procedures based on pedo-transfer functions. The soils database was updated for each of the soils to be modelled with basic soils information - content of sand, silt, clay and organic carbon, bulk density, pH and description of soil structure. The system then automatically calculates all the soils properties required as input for MACRO and these were used without any changes at all to simulate observed behaviour.

5.3 RESULTS

Results for simulation of the four datasets with MACRO and MACRO_DB are summarised in Tables 1-4.

Table 1 *Comparison between maximum concentrations of isoproturon (μg/l) observed in six events at Brimstone Farm and those simulated in uncalibrated runs with the models*

Season	Event	Observed*	MACRO	MACRO_DB
1	First event	465 (132)	69.1	0
	Second event	134 (47)	55.7	0
2	First event	65.1 (14.7)	566	8.2
	Second event	2.6 (2.3)	524	50.1
3	First event	0.64 (0.41)	58.5	35.1
	Second event	0.21 (0.28)	12.4	2.9

* Mean of values for four plots, standard deviation in parentheses

Table 2 Comparison between maximum concentrations of isoproturon (μg/l) in two events monitored at Wytham and those simulated in uncalibrated runs with the models

Event	Observed	MACRO	MACRO_DB
First event	290	9	0
Third event	129	135	73

Table 3 Comparison between total losses and maximum concentrations of isoproturon (IPU) at Cockle Park and those simulated in uncalibrated runs with the models

Parameter	Observed	MACRO	MACRO_DB
Loss of IPU (mg/m^2)	0.141	1.53	0.0004
Max. conc. of IPU (μg/l)	4.2	13.1	0.14

Table 4 Comparison between observed total loss of isoproturon (mg m^{-2}) from the SSLRC lysimeters and that simulated in uncalibrated runs with the models

Soil type	Mean observed	MACRO	MACRO_DB
Unstructured sand	0.28	1.00	1.74
Loam over gravel	0.92	0.52	0
Deep loam	3.24	4.91	0
Alluvial clay	3.91	1.33	0
Peat over sand	0	0	0

5.3.1 MACRO

MACRO showed considerable variability in predictive ability, but generally the model performed better than the non-preferential flow model LEACHP even for sandier soils, where preferential flow is of minor importance. At Brimstone Farm, MACRO under-estimated maximum concentrations of isoproturon in the first and over-estimated them in the remaining two seasons (Table 1). On the other heavy clay soil at Wytham, an excellent simulation was obtained for the third event after application although both volumes of drainflow and pesticide concentrations for the first event were under-estimated (Table 2). The simulation for Cockle Park gave a good match to initial concentrations of isoproturon, but enhanced movement late in the season relative to that observed suggests a problem with over-prediction of pesticide movement in matrix flow over longer periods of time (Figure 3).

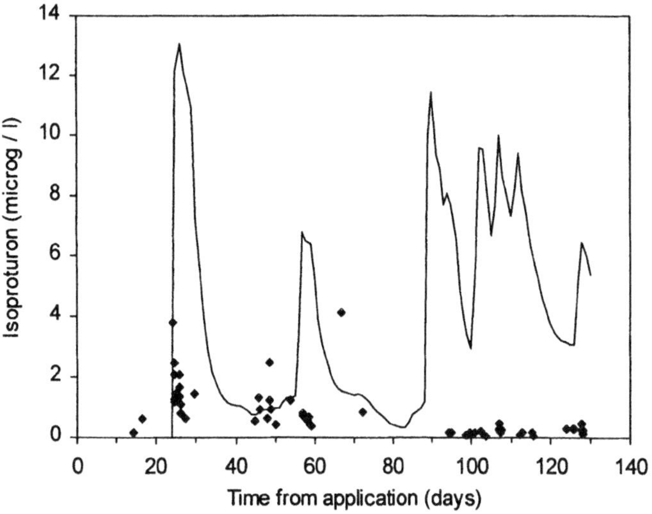

Figure 3 *Comparison between observed concentrations of isoproturon in mole drainflow from Cockle Park (♦) and those simulated by MACRO (____)*

Results for the intermediate soils in the SSLRC lysimeter dataset were relatively encouraging with uncalibrated simulations giving reasonable estimates of total isoproturon losses from all five soils (Table 4).

Levels of predictive accuracy of MACRO modelling for the four datasets are summarised in Table 5.

Table 5 *Levels of predictive accuracy for uncalibrated simulations of the four datasets with MACRO (all values are predicted values as a factor of the observed; maximum concentrations are for the whole simulation and take no account of timing of maximum)*

Dataset	Maximum pesticide concentration	Total loss of pesticide
Brimstone Farm	0.15-91.4	-
Wytham	0.47	-
Cockle Park	3.1	10.9
SSLRC lysimeters	0.26-1.84	0.34-3.57

The levels of predictive accuracy presented should not be taken as general values and it should be considered that the number of datasets was limited and concentrated almost exclusively on one compound (isoproturon)[1]; the datasets were of a generally high quality; the modellers involved were relatively experienced with MACRO. MACRO predicted simulated leaching from a range of soils at an acceptable degree of accuracy. Predictive ability was greatest for the longer simulations and the coarser-textured soils.

5.3.2 MACRO_DB

Relative to the soil parameters used with the stand-alone version of MACRO, those automatically selected within MACRO_DB reduced the emphasis on preferential flow for all of the simulations. This was because the boundary between micropores and macropores was set closer to saturation using pedotransfer functions than using expert judgement, even though site-specific soils data were used within MACRO_DB. Perhaps more importantly, sorption in the macropore region was assumed to be larger in simulations with MACRO_DB than with MACRO. As a result, MACRO_DB simulated lower concentrations of pesticide than MACRO for all model runs (Tables 1-4, Figure 4). Generally, this decreased the accuracy of the pesticide simulation, although there were exceptions where the lower concentrations were closer to those observed (Brimstone Farm in the third season). In a number of cases MACRO_DB failed to predict any loss of pesticide to drains from the clay soils (first season at Brimstone and first event at Wytham) when high concentrations were actually observed, whilst losses from the clay loam soil at Cockle Park were greatly under-estimated (Figure 4). For the intermediate soils studied in the SSLRC lysimeters, MACRO_DB simulated no leaching of isoproturon from four of the five soils whereas only one of the soils was observed to show no consistent leaching over the experimental period.

[1] The models were also evaluated against concentrations of trifluralin in drainflow at Cockle Park. No losses were simulated although consistent small losses were observed [22]. These results are not included in Table 5

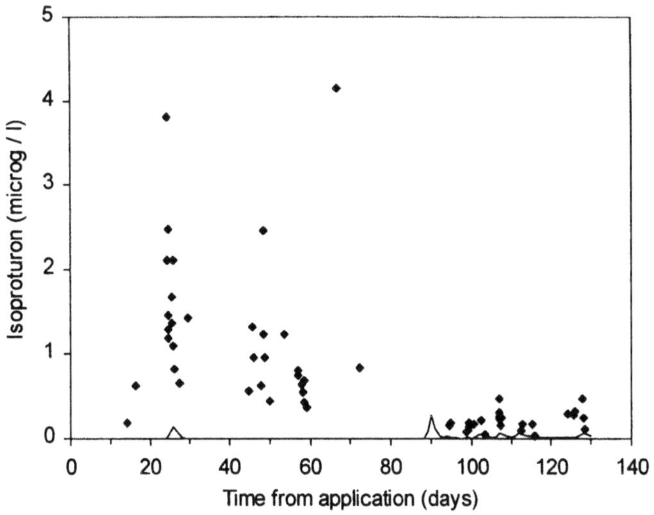

Figure 4 *Comparison between observed concentrations of isoproturon in mole drainflow from Cockle Park (◆) and those simulated by MACRO_DB (——)*

5.4 CONCLUSIONS

On balance, the findings of the evaluation by Beulke *et al.* [22] suggested that MACRO should continue to be the preferred preferential flow model for regulatory purposes. Its predictive ability was good in a range of intermediate soils, less good in the clay loam and very variable in the two heavy clay soils where there are inherent difficulties in predicting observed behaviour because of the extreme spatial and temporal heterogeneity in their structure. The model is user-friendly and well-documented. Nonetheless, it was concluded that at the present time, MACRO should be used for regulatory purposes with caution. Parameter selection for MACRO is still problematic and the model should only be applied by an experienced user. A comprehensive calibration step should be included wherever possible. A useful development for the stand-alone version of MACRO would be some general guidance on realistic values for these parameters in a range of representative soils.

The philosophy behind MACRO_DB to automatically select parameters using estimation algorithms is commendable. However, the consistent under-estimation of preferential flow relative to matrix flow in a broad range of soils suggests that further work on parameter selection and extensive testing against field data are required before the system can be considered valid as a regulatory tool.

6. Outlook

It is likely that MACRO will continue to be the preferred preferential flow model for regulatory use within Europe for the foreseeable future. Current misgivings over the accuracy and robustness of models are likely to diminish as more results of validation studies become available. The development of more guidance on parameter selection for users of the stand-alone version of MACRO and further work on MACRO_DB will result in user-friendly systems, which facilitate robust parameter estimation. This will, in turn, increase confidence in modelling results. Nonetheless, the use of MACRO to simulate pesticide leaching to groundwater is likely to be restricted to higher tier modelling within national registration and its use on a European level will continue to be limited. This is partly due to the fact that finer-textured soils for which preferential flow is important are often not overlying major aquifers. In contrast, preferential flow models are expected to be assigned an increasing role in risk assessments for pesticide losses to surface waters. The large transient concentrations in surface waters resulting from pesticide losses via preferential flow are considered to be relevant from an ecotoxicological viewpoint and these pulses cannot be simulated with conventional models.

Given the extreme heterogeneity of soils which are prone to preferential flow, deterministic modelling using a single set of input parameters cannot give a realistic representation of the range of concentrations which can be observed in drainflow or leachate from a single site, although the mean behaviour (*i.e.* spatially and/or temporally integrated) can be simulated. It is desirable that the importance of probabilistic or stochastic modelling which takes the variability of influencing parameters into account will increase. However, the routine use of such approaches in regulatory risk assessments is hampered by relatively large requirements of data to define distributions of model parameters and by complex inter-correlations between parameters.

7. Acknowledgements

The study presented in Section 5 was funded within the UK Pesticides Research Programme of the Ministry of Agriculture, Fisheries and Food. Data for model evaluation were provided by ADAS, IACR-Rothamsted, University of Newcastle, Institute of Hydrology and Horticulture Research International (UK).

8. References

1. BBA, (1990). Guidelines for the testing of plant protection products in registration procedures. Part IV, 4-3: Lysimeter tests for the translocation of plant protection products into the subsoil. Biologische Bundesanstalt, Braunschweig, Germany.

2. Berg, F. van den & Boesten, J.J.T.I. (1998). Pesticide leaching and accumulation model (PESTLA) version 3.3. Description and user's guide, Technical Document 43, DLO-Staring Centrum, Wageningen, The Netherlands.

3. Brown, C.D., Hollis, J.M., Bettinson, R.J., Beulke, S. & Fryer, C.J. (1997). Pesticide Mobility: Lysimeter Study to Validate the Relative Leaching Potential of UK Soils. Research Report for MAFF Project PL0510.

4. Beven, K. & Germann, B. (1982). Macropores and water flow in soils.*Water Resour. Res.*, **18**, 1311-1325.

5. Harris, G.L., Nicholls, P.H., Bailey, S.W., Howse, K.R. & Mason, D.J. 1994. Factors influencing the loss of pesticides in drainage from a cracking clay soil. *Journal of Hydrology*, **159**, 235-253

6. Johnson, A.C., Haria, A.H., Bhardwaj, C.L., Völkner, C., Batchelor, C.H. & Walker A. (1994). Water movement and isoproturon behaviour in a drained heavy clay soil: 2. Persistence and transport. *Journal of Hydrology*, **163**, 217-231.

7. Brown, C.D., Hodgkinson, R.A., Rose, D.A., Syers, J.K. & Wilcockson, S.J. (1995). Movement of pesticides to surface waters from a heavy clay soil. *Pesticide Science*, **43**, 131-140.

8. Flury, M., Leuenberger, J., Studer, B & Fluhler, H. (1995). Transport of anions and herbicides in a loamy and a sandy field soil. *Water Resources Research*, **31**, 823-835.

9. Aderhold, D. & Nordmeyer, H. (1995). Leaching of herbicides in soil macropores as a possible reason for groundwater contamination. In: *Pesticide Movement to Water* (eds. A.Walker *et al.*), BCPC Monograph No. 62, British Crop Protection Council, Farnham, Surrey, 217-222.

10. Flury, M., Flühler, H., Jury, W.A. & Leuenberger, J. (1994). Susceptibility of soils to preferential flow of water: a field study. *Water Resour. Res.*, **30**, 1945-54

11. Armstong, A.C., Matthews, A.M., Portwood, A.M. & Jarvis, N.J. (1995). CRACK-NP. A model to predict the movement of water and solutes from cracking clay soils. Version 1.0. Technical description and user's guide. ADAS Land Research Centre, Gleadthorpe, Notts.

12. Hall, D.G.M. (1993). An amended functional leaching model applicable to structured soils. I. Model description. *Journal of Soil Science*, **44**, 579-588.

13. Jarvis, N.J. (1994). The MACRO model (Version 3.1). Technical description and sample simulations. Reports & Dissertations 19, Department of Soil Science, Swedish University of Agricultural Sciences, Uppsala.

14. Ahuja, L.R., DeCoursey, D.G., Barnes, B.B. & Rojas, K.W. (1993). Characteristics of macropore transport studied with the ARS Root Zone Water Quality model. *Transactions of the ASAE*, **36**, 369-380.

15. Brown, C.D. & Hollis, J.M. (1996). SWAT - A semi-empirical model to predict concentrations of pesticides entering surface waters from agricultural land. *Pesticide Science*, **47**, 41-50.

16. Boesten, J., Businelli, M., Delmas, A., Edwards, V., Helweg, A., Jones, R., Klein, M., Kloskowski, R., Layton, R., Marcher, S., Schäfer, H., Smeets, L., Styzcen, M., Russell, M., Travis, K., Walker, A. & Yon, D. (1995). Leaching models and EU registration. The final report of the work of the Regulatory Modelling Work group of FOCUS (FOrum for the Co-ordination of pesticide fate models and their Use), EU Doc. 4952/VI/95, 123 pages.

17. Adriaanse, P. Allen, R., Gouy, V., Hollis, J., Hosang, J., Jarvis, N., Jarvis, T., Klein, M., Layton, R., Linders, J., Schäfer, H., Smeets & L Yon, D. (1996). Surface water models and EU registration of plant protection products. Final report of the work of the Regulatory Modelling Work group on Surface Water Models of FOCUS (FOrum for the Co-ordination of pesticide fate models and their USe), EU Doc. 6476/VI/96, 227 pages.

18. Jarvis, N.J., Hollis, J.M., Nicholls, P.H., Mayr, T. & Evans, S.P. (1997). MACRO_DB: a decision support tool for assessing pesticide fate and mobility in soils. *Environmental Software*, **12**, 251-265.

19. Nicholls, P.H. (1994). "Physicochemical evaluation: the environment", an expert system for pesticide preregistration assessment. Proceedings of the Brighton Crop Protection Conference - Pests and Diseases, 1337-1342.

20. Hollis, J.M., Hallett, S.H. & Keay, C.A. (1993). The development and application of an integrated database for modelling the environmental fate of herbicides.*Proceedings Brighton Crop Protection Conference - Weeds - 1993*, **3**, 1355-1364.

21. FOCUS (2000). FOCUS groundwater scenarios in the EU pesticide registration process. Report of the FOCUS Groundwater Scenarios Workgroup, in preparation.

22. Beulke, S., Brown, C.D., & Dubus, I. (1998). Evaluation of the use of preferential flow models to predict the movement of pesticides to water sources under UK conditions. Research report for MAFF project PL0516, Soil Survey and Land Research Centre, Silsoe, Beds, UK, 101 pp.

23. Hutson J.L. & Wagenet R.J. (1992). Leaching Estimation and CHemistry Model, Version 3. Dept. of soil, crop and atmospheric sciences, Research series No. 92-3. Cornell University, New York.

A SCIENTIFIC AND TECHNOLOGICAL FRAMEWORK FOR EVALUATING COMPARATIVE RISK IN ECOLOGICAL RISK ASSESSMENTS

JOHN M. JOHNSTON
Ecosystems Research Division,
960 College Station Rd.,
Athens, GA 30605
USA
johnston.johnm@epa.gov, voice (706)355-8153, fax (706) 355-8104

Abstract

There are significant scientific and technological challenges to managing natural resources. Data needs are cited as an obvious limitation, but there exist more fundamental scientific issues. What is still needed is a method of comparing management strategies based on projected impacts to ecosystem health. Ecological risk assessment is a field in its infancy, and its focus has been primarily toxic hazards (i.e. pesticides) to aquatic endpoints. Expanding on these achievements with the expression of sustainable, edible fisheries in an entire estuary as an assessment endpoint, and with greater complexity than a single species or species-by-species approach, is a first challenge. The extension of the scope of a risk assessment to include non-chemical stresses, such as land use change and nitrogen enrichment, is requisite to managing resources given the significance of how these disturbances alter hydrologic balances, habitat characteristics, and even the structure of ecological communities. The separation of intrinsic variability in the status of the fisheries from those variations that result from anthropogenic sources of disturbance is also a challenge that is not trivial. Management alternatives are thus evaluated based on the costs of remediation and related economic and societal issues and the projected changes in resource quality. Ultimately, terrestrial endpoints require attention as well. As an interdisciplinary application of such fields as ecology, biology, environmental management, toxicology, hydrology, and economics, ecological risk assessment requires a much broader, more comprehensive scope and a conceptual framework that synthesises the contributions of the supporting science and management.
These challenges combine with the practical, technological challenges of how to conduct a risk assessment. Central to the goal of performing analyses of various resource management scenarios is the need for a computer-based problem solving environment that automates many of the associated tasks: data gathering and manipulation, integration of statistical, empirical, and mathematical simulation modelling and analysis techniques, and the accommodation of model inter-comparisons within a common framework. Because there are no rules as such for performing an ecological risk assessment,

J.B.H.J. Linders (ed.), Modelling of Environmental Chemical Exposure and Risk, 133–150.

the guidelines that exist as expert knowledge could also be codified and made available within such a framework. It is important to understand that such a framework is much more than simply a collection of assorted tools in a software toolkit. It is the implementation of the science for performing comparative ecological risk. Advances in ecological risk assessment are of a scientific as well as technological nature, and any hoped for state-of-the-art applications of the field must eventually give attention to both areas of need. I present the ongoing development of both a scientific conceptual model for performing comparative risk and a software framework to meet these needs.

1. Introduction

Environmental protection can be categorised by two modes: *restorative*, when problems are designated as deserving of human attention and action, and *anticipatory*, such as when the environmental outcomes of human actions are to be assessed for management or regulatory purposes. Risk assessment and risk management have been suggested as a problem solving framework for addressing the general needs of both environmental and human health protection [14]. However, because risk assessment has been applied mostly to improving the condition of specific endpoints or evaluating the risks associated with a specific kind of stress or hazard (e.g., evaluating risks of chemical releases to an aquatic population) it remains to be seen as to whether risk analysis is suitable for broad protection of ecosystems. Risk managers seek to reduce or eliminate risk posed by an identified hazard to a clearly described endpoint. Thus the restorative use of risk assessment is apparent as well as the limited predictive use for focused stressor / receptor analysis. Such an approach is daunting though if the predictive scope is widened to multiple stresses and receptors, and the solutions are even more problematic when a host of risks overlap and interact. Simply put, the motivation and focus of the risk assessment / risk management framework is both its strength and its weakness.

Decreases in uncertainty are achieved by limiting the scope of the question or problem, but this reduction comes with a significant cost. Environmental protection is unlikely to be achieved through the management of even a collection of well-selected endpoints (native biodiversity, mass mortality are examples), since systems are too complex for such a piecemeal approach. What is needed is a means of fully treating the system of interest as a functioning whole, encompassing both human activities (and the alteration of the environment) and the ecological resources that exist. I propose that sustainability assessment is such a means, and that a complete system can be addressed before making assumptions about primary causes of stress and the most important pathways and endpoints of concern. Using a combination of theoretical and empirical models, the cause and effect relationships in an ecosystem are described by a series of linked models, and scenarios are developed that incorporate a range of possible future climate and human environmental use changes and the relating ecosystem states. Sustainability is then defined as the difference in the comparison of virtually any modelled state variable in present and predicted alternative future states. Examples include identifying where species, guilds or communities disappear from their current ranges or where fish populations have contaminant residue concentrations in their tissues at levels that lead to human health concerns and fish mortality.

A prediction of system sustainability is both scientifically meaningful and a tractable management endpoint, and it represents the characterisation of all risks to the system rather than a chemical-by-chemical or media specific risk assessment. For example, considering that various fish population and community dynamics will be predicted via computer simulation, missing size classes as well as pesticide tissue concentrations can be evaluated within the assessment based on any number of related environmental and human factors– fish harvesting, pesticide use, climate change, degree of urban and sub-urban development. These processes are incorporated into the sustainability assessment itself as the mechanisms connecting human activities to aquatic ecosystem dynamics. To truly protect and safeguard the environment we must develop a means of anticipating where problems will occur-- instead of waiting for them to develop before taking action. The results of a sustainability assessment are tractable management endpoints in that the risks are expressed as events per unit time and not probability estimates. Examples of this include the number of extinctions predicted to occur over a 50-year period, the decrease in types of habitat in a region or the reduction in edible sport fish from current levels. These findings clearly communicate the nature of the risk, which is the ultimate goal of any effective risk assessment for decision-making [14].

Currently we lack the technical tools for conducting such an assessment, given that we need to bring many separate science components together to simulate human activities and ecosystem response. Connecting models of various kinds is no trivial task, considering that most models are not designed to be used as components or to communicate there are a number of data I/O and interface issues that are apparent. Johnston *et al.* [11] have outlined a vision for the development of a modelling framework that will support such a sustainability assessment. It is the long-term development of a Multimedia Integrated Modelling System (MIMS) that is needed to effectively and flexibly utilise a collection of science components for environmental research and management.

2. Sustainability Assessment

The development of risk assessment methods to hazards that go beyond aquatic ecosystems and chemical hazards has not been rapid. In fact, success for the extension of risk assessment to non-chemical threats is most likely when the environmental hazards are analogous to chemical releases. The risk posed by an exotic species introduction is such an example. As Suter (p. 393, [13]) writes, “For chemicals, choosing endpoints is largely a matter of determining what valued environmental components could be significantly exposed to the chemical. For organisms, choosing endpoints is largely a matter of examining the traits of the organisms to determine what it might do in the ambient environment and then inferring what valued components of the environment might be affected by those activities.”

This example is also useful in reinforcing the point concerning the limitations of the component-by-component approach that is the hallmark of a good risk assessment: make a list of organism traits and then identify those endpoints that could be affected directly by the activities of the introduced biological hazard. Indirect effects of a hazard are not of immediate interest, and although one may be interested in the value of the system and its functioning wholeness in practice the scope of the assessment is limited to the response of the isolated components to (typically chemical) exposures. The shape of the exposure-response function is of particular interest in completing a risk assessment, and it represents a critical piece of the analysis. However, the dimensions of ecosystem response are perhaps never to be simplified to a two dimensional graph.

Can the cause and effect relationship for all interactions be expressed as an exposure / response curve, such that a mesocosm toxicity test of varying dose exposures is of ecosystem response in a general sense? What is implied by this approach is that all related dependent variables can be described with multi-dimensional exposure curves. It is also implied that daily intake rates, reference doses or reference concentrations can describe all risks of interest. In practice we are interested in stresses that are not analogous to dermal contact, inhalation or ingestion. There is also a semantic difficulty in that exposure is defined as the condition of being exposed to something detrimental, a condition of being unprotected and subject to some effect or influence. Does it follow that ecosystems are *exposed* to land use change, climate change, and related factors such as nutrient enrichment, habitat loss and fragmentation? Such a term seems inadequate when applied in this manner. The language of risk assessment is that of a system that is affected by a perturbation, and the perturbations considered are usually that of introduced pollutants (toxic industrial waste, municipal sewage, pesticides) that stress ecosystems. Are all risks equal and comparable? It is suggested that land use change and hydrologic alteration are the greatest sources of environmental risk to our biological resources, even though other risks are known exist (e.g., contaminated sediments, ozone depletion or acid deposition) [12].

The items listed in Table 1 are examples of what can be considered ecosystem stresses and the related features of an organism's habitat that are the direct exposures of changes in land use and climate. These can be considered graphically in Figure 1 and Figure 2, which taken together describe lethal and sublethal ecosystem responses to perturbations. Such a presentation in enlightening, since more of the complexity is conveyed in this manner than possible in a list or tally. These are the variables that must be considered in an aquatic sustainability assessment. Figure 3 depicts the general approach of conducting a sustainability assessment, utilising a combination of models for various purposes to complete the ecological assessment. The scenarios are those variables that change during the period of the assessment, the driving variables considered here are climate and human land use. Properties such as soil type remain constant for the duration of the assessment, as do the processes describing the hydrology, hydrodynamics and

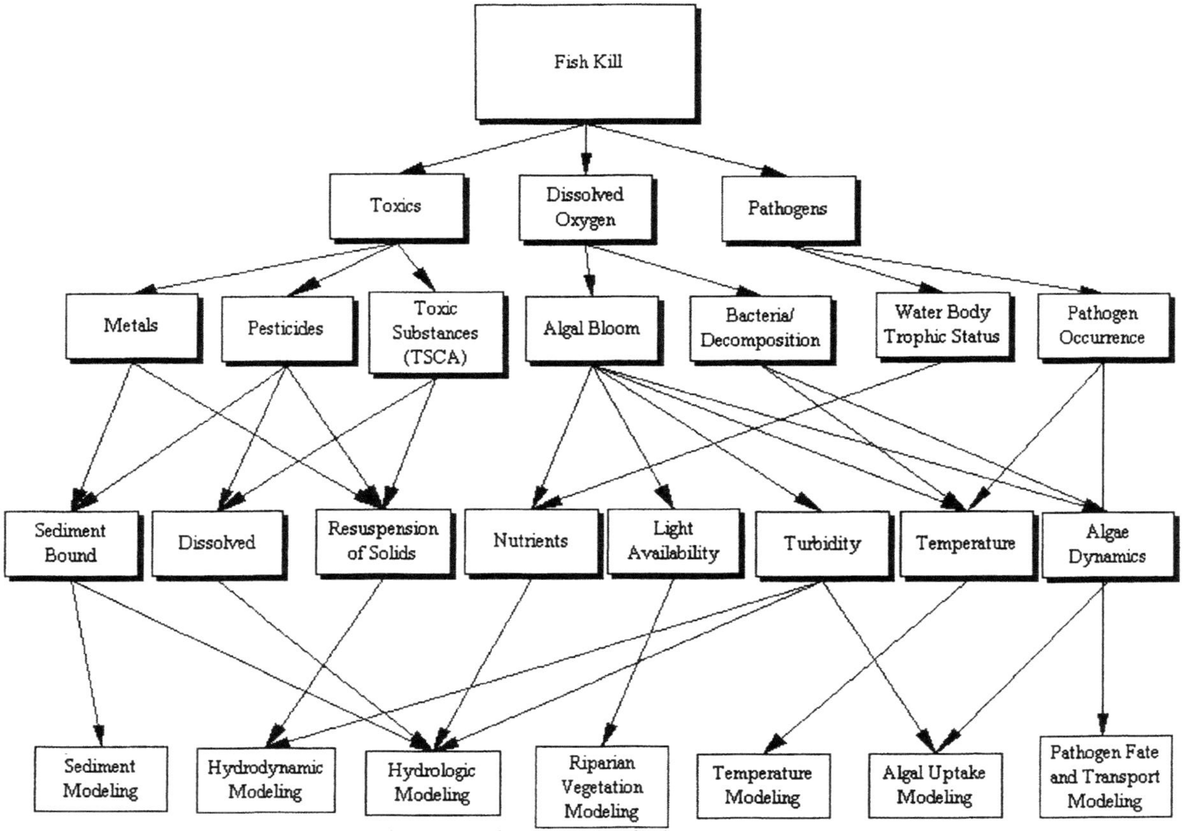

***Figure 1**. Fish health as an acute biological response of a system to multiple stressors.*
Included in this diagram are the various causative factors and related variables for presence / occurrence. In the bottom row are possible modelling categories for describing the variables and their dynamics

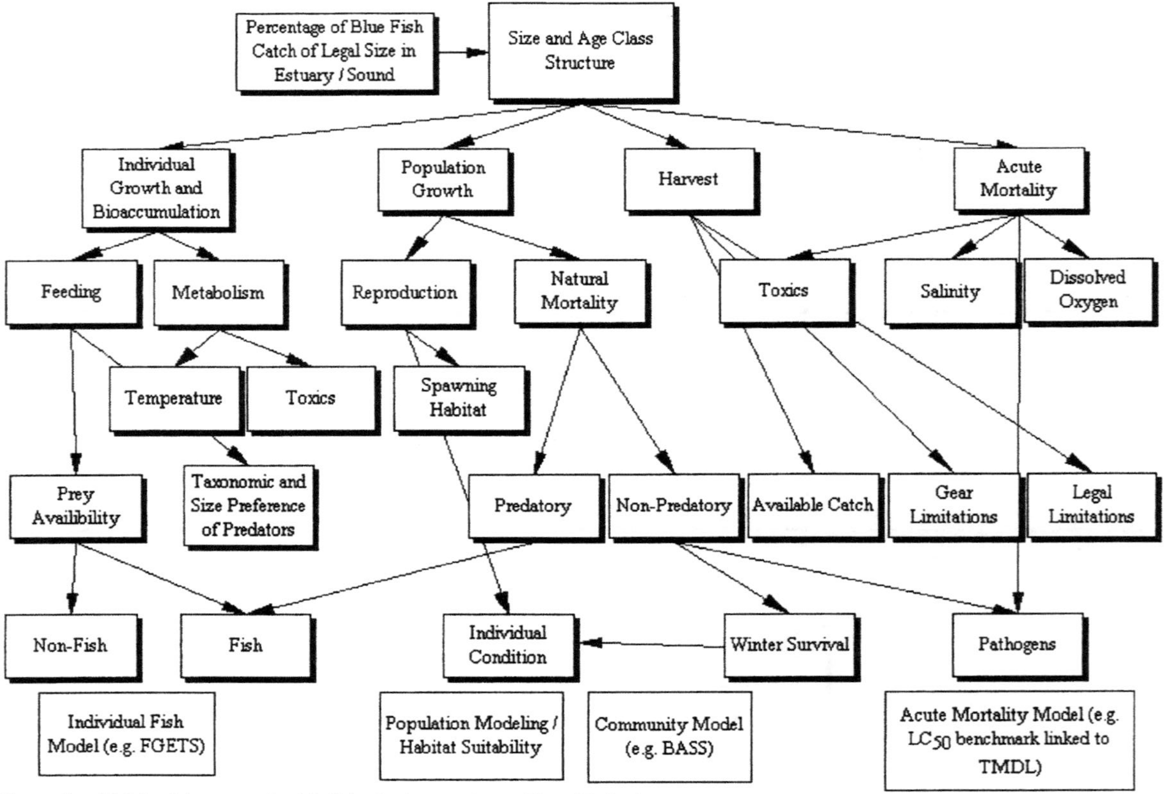

Figure 2 *Fish health as sustainable fisheries in a system with multiple stressors.*
This diagram includes the previous ("acute mortality" and related) as well as the more complete description of variables in the system that include individual condition, population and community interactions, and human influences via harvesting. The categories of modelling, with examples, are provided to illustrate the treatment of these variables

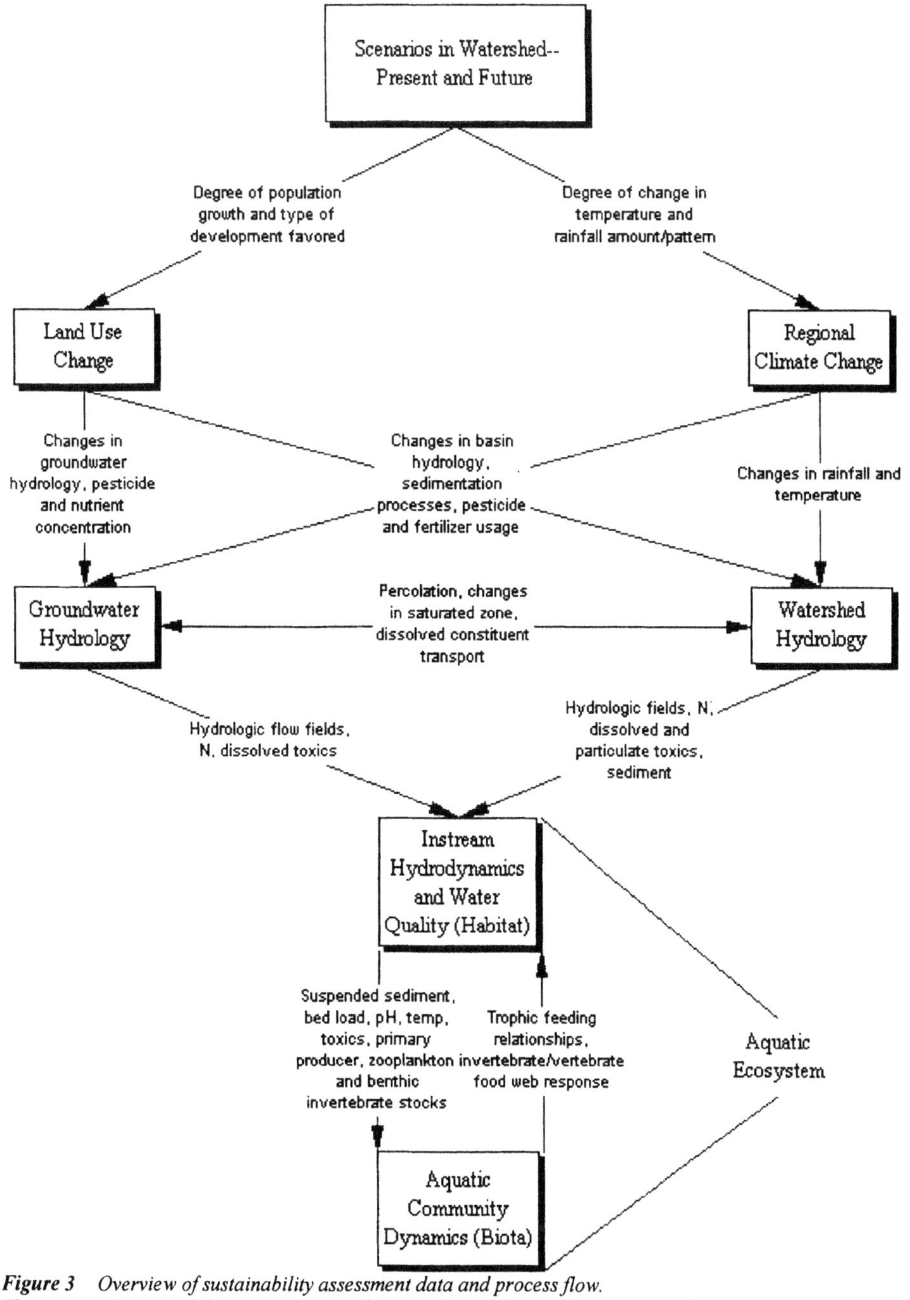

Figure 3 *Overview of sustainability assessment data and process flow.*
The proposed sustainability assessment is illustrated as a linkage of various models for system description, as well as the means of establishing mechanisms for human changes of the watershed and instream responses. The aquatic community is depicted as a dynamic feedback between higher order trophic interactions and a traditional water quality model (for hydrodynamics, water quality, and sediments).

chemistry. Similar to risk assessment, the essence of sustainability assessment is comparative. This comparison of present to future alternative states is shown in Figure 4.

TABLE 1. Categories of Stress and Proximal Exposure Factors for Instream Biota

Land Use Change	Climate Change
Suspended sediments	
Bottom sediments	
Pesticides	
Increased nitrogen / phosphorous	
Increased carbon compounds (via municipal waste)	increased carbon compounds (via increased plant production)
Heavy metals	
Frequency and amount of water movement (via channels, impervious surfaces)	frequency and amount of water movement (via rainfall changes)
Barriers to movement (via dams)	
Exotic invasion (via introduction)	exotic invasion (via range changes)
Temperature change (via changes in stream side-vegetation, cold-water dam release)	temperature change (via expected warming)

Since the modelling system that comprises the mechanics of the sustainability assessment is a mix of empirical and mathematical models, it will not be possible to simulate the combined watershed / human / ecosystem behaviour dynamically for an assessment period, whether 5, 10, or 50 years into the future. Instead, the approach is to duplicate both current conditions and future conditions for a period of one to a few years duration. After setting initial conditions, models will be run to achieve a nominal or equilibrium behaviour to simulate present and future watershed dynamics for each initial and final time period. It is the difference in all modelled states, whether by season, annual average, or dynamics as produced for the simulation of a few "typical" years that provides information about sustainability. Although the flow of information in this model approach is largely one way, coupling is possible at any point if desired if it is necessary to reproduce desired system behaviour. Since the aquatic ecosystem is of interest in a holistic sense, a coupling of instream water quality to the trophic-dynamic models is proposed (as shown in Figure 3). Since there are known to be interactions between benthic invertebrates, organic materials, zooplankton and primary producers, it is important that the system conceptual model reflects this as much as possible.

The interesting, and possibly more controversial aspect of the proposed approach involves the creation of the driving scenarios. A description of the current status involves knowledge of human population density and types of land use, of which the latter is available from remotely sensed datasets. From these data we will infer amounts

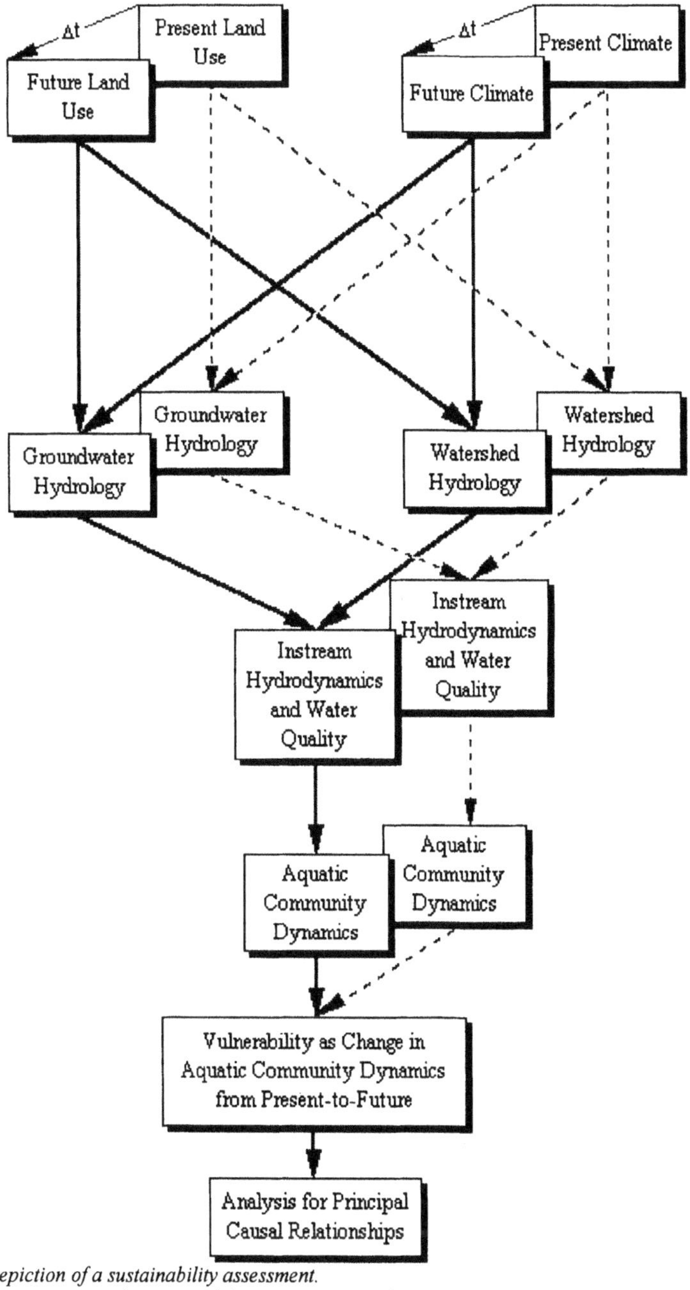

***Figure 4**. Depiction of a sustainability assessment.*
The comparative essence of a sustainability assessment is shown. Sustainability is expressed as the change in an aquatic state variable, present to future. The driving variables, those which change in time as between alternative futures, are land use (nutrients, pesticides, sediments) and climate.

of pesticide and fertiliser use, as well as other aspects of agricultural land use that affect nutrient balances (animal feeding operations, livestock grazing, etc.). Many assumptions will of necessity be made when estimating future human population densities, land use and agricultural practices, and climate change. A range of scenarios will be constructed to show the extremes in human population change and ecosystem response. In time more variations will be added to the analysis so that it is more likely that a particular risk manager will be able to identify with a particular scenario.

The output of the analysis is a database of all model variables for both initial and final assessment times. Since it is not obvious what is needed for investigating causal relationships, all information will be available to the risk assessor / manager. Visualisation of any variable of interest, provided as model output, will be possible: this includes age and size classes of fish populations, species abundance, population mortality, and annual (or monthly, daily, or seasonal) patterns of sedimentation, dissolved oxygen levels, water temperature, and eutrophication. Because the predictions are geographically referenced to watersheds of various drainage areas, it will also be possible to visualise changes in various ecosystem resources in time in a spatially explicit manner.

A means of performing the sustainability assessment is not obvious. A likely starting place is that of co-ordinating the participation of the necessary model experts for each domain: land use change, watershed hydrology, hydrodynamic and instream water quality modelling. Moving information between the modelling domains is a challenge, since there are differences in spatial and temporal resolution in any collection of models never intended to be linked as proposed. Model parameterisation and input data formatting will be necessary, as will output post-processing and visualisation. The need for a comprehensive assessment of the physical, chemical and biological components of a system is apparent. The task of ecological risk assessments demands that we construct ever more complicated models (multi-media, multi-pathway, multi-stressor, and multi-receptor) such that a rather sophisticated and complex ecosystem model is produced. How to effectively manage and test such a model is a significant challenge for those engaged in environmental protection. As illustrated in Figures 1 and 2, there are many interrelated processes and variables that could be treated separately, though typically these are combined to create larger abstractions of the system, as population and community models, watershed hydrology and fate models, and exposure models. Although it is highly desirable to have the flexibility of integrating the actual process descriptions of interest, at the present this is not possible and instead the aggregated models are used for the science they contain.

3. Multimedia Integrated Modelling

In order to perform a complete sustainability assessment we would ideally like to have a modelling system capable of selecting and linking atmospheric, hydrologic, hydrodynamic, and aquatic population and community models into a consistent computational and execution framework. At present this is not possible. Although there is a collection of existing models that could potentially be used for the assessment, there are significant technical and scientific difficulties in such a strategy. In addition to the obvious challenge of using models in a way that they were never designed, the use of legacy models that were created to represent a "model system" also have the following disadvantages: 1) boundary simplifications due to the lack of scientific expertise of the developer, 2) assumptions and simplifications necessary to complete a system model when some descriptions were originally of primary interest to the developer (uneven level of rigor in detail and approach), 3) greater data, parameter, and related formatting effort, and 4) the fact that the model represents the developer's "representative case" of the modelled system in consideration.

There is the need for an overall modelling and decision support tool that integrates the execution of various environmental models, so that sources of stress can be linked via mechanistic cause-and-effect connections to the physical, chemical and biological endpoints of interest. Such a linking of environmental codes is unprecedented in a dynamic sense (with feedbacks where they exist), and there is general acceptance of the need to tackle multimedia problems in a holistic and comprehensive manner as possible. The recent direction of the National Exposure Research Laboratory (NERL) of the Office of Research and Development (ORD) is to produce a flexible multimedia-modelling framework to meet a variety of regulatory needs. Such a framework has been designated MIMS, the Multimedia Integrated Modelling System, and represents an ambitious, long-term research undertaking on the part of NERL.

The recognition of the need for a more flexible, comprehensive modelling approach for performing risk assessments is not new even within the US EPA. As investigators of the U.S. EPA EcoRisk Program, Barber [3] discusses the appropriateness and adequacy of empirical and theoretical-based approaches and their applicability in risk assessments, and Bird [6] addresses the application of models originally developed for the chemical exposure of fish to the assessment of avian endpoints. Bird presents a description of the modular format of the terrestrial exposure system, PIRANHA [8], adding that the component construction of the system provides for future integration of updated model versions. However, it is shown in this example that an even greater degree of integration between components and greater flexibility is desired than was proposed by the assembled models. In the case of PIRANHA it was uncertain which spray drift model to select for the simulation of aerial pesticide application, and with two candidate models created for different purposes (differing in data requirements as well as model structure) there was no equivocal reason to include or exclude either. However, it was not possible to select both and then use each as best fit a particular application, since this single model would be linked to other single model selections for the risk assessment framework [6]. The other components of the pesticide risk assessment framework included: PRZM [9] for soil pesticide movement, EXAMS [7] for calculating pesticide concentration in streams and ponds, EPIC [15] to simulate crop growth and pesticide uptake, and FGETS [4] for estimating pesticide accumulation in terrestrial fauna. Bird [6] discusses the uncertainty of the proposed framework, considering, for example, that FGETS was developed to simulate accumulation of neutral organics in fish and EXAMS was designed for pesticide fate and transport in streams and ponds, as well as the difficulties in validating such a risk assessment framework.

Because the EcoRisk program proposed a design for a risk program that was fixed in its science components and application context, many of the issues of flexible model integration were avoided. Namely, it was never intended that other candidate models be evaluated for later inclusion, or that the proposed framework would assist the risk assessor to evaluate or even improve the component science codes. Although Bird [6] addresses the significant challenges of data management and model input for such a system, an assessment of the framework itself is not really an issue since the models contained within are essentially independent and have undergone separate peer reviews. But what of the science contained in these models, which likely makes up a much smaller portion of the software involved? Barber [2] questions the validity of the assumption that the primary route of pesticide exposure for birds is via food chain transfer. Unfortunately there is no way to test the validity of this assumption using the risk framework, since the framework itself imposes that assumption. It seems reasonable to expect that there will be a need for a much greater degree of scrutiny in a combined framework, and this includes close scrutiny of model assumptions, the use of empirical, statistical, and theoretical/mathematical methods, temporal and spatial resolutions, and the quadrature of model solution methods (to name just a few). How to go about this is not straightforward, and a significant difficulty is that models are not typically created and documented in such a way as to be easily reviewed, improved, and reused. The intelligent use of any science component is limited by any of the information a scientist (or software interface) might need to know about the formulation and intended use of science process descriptions. This science model metadata is independent of the implementation of the science and must be available in order to make a framework approach successful.

A rigorous approach to model documentation has yet to be achieved, though it has been noted and attempted recently [10]. Hoch *et al.* have created an information system for both models and mathematic process descriptions accessible through the Internet. Their goal is to provide a central, standardised database for (primarily) ecological process descriptions, so that the larger scientific community can have improved access to this information for, in essence, international peer review and continued refinement. These authors firmly believe that for models to be truly useful, they must be well documented and testable by the scientific community to the level of model subcomponents [10]. By dividing large models an evaluation of the subcomponents will reveal possible shared patterns, as well as those components suitable for use in building new models as needed. Such an analysis, if made available to more than simply a limited few who are intimately involved with a particular model formulation, will also reduce potential model misapplication and enable a thorough peer review. Given the current manner that models are typically published and tested via parameter sensitivity analysis, a thorough evaluation of a large and complex model is a nearly impossible task [5]. The only acceptable goal is that of making model assumptions and process algorithms transparent to potential model appliers and developers. Only when this is achieved models will be properly peer reviewed and trusted for the knowledge they contain. A more open disclosure and democratic process of testing and refinement is the hallmark of an effective scientific process.

Making use of this well-documented repository of scientific environmental process understanding is another significant challenge. It is this challenge in particular that MIMS will undertake to the benefit of both environmental researchers and the management community. Hoch *et al.* [10] mention that they intend to pursue a means of translating the model documentation to simulation languages for use in creating models from available subcomponents. MIMS is intended as a software framework to provide this functionality, and as such represents a dramatic change in the manner in which environmental models are developed and understood. Not that modularity is a new idea in the construction of environmental models, as "Modularity in Plant Models" was a special issue topic of Ecological Modelling [1].

MIMS science components will be selected from a thorough literature review, beginning with those process descriptions of priority in describing aquatic ecosystems, including small and large flowing freshwater systems, lakes and reservoirs, and estuaries and sounds. An example of a process description that will be requisite in the MIMS is that of the exchange of toxic substances with fish via gill and food intake, taken from the FGETS model [4]:

$$d\frac{B_T}{dt} = Sk_wC_w + C_FF - \frac{B_T}{W_TBCF}\left[Sk_w + Ek_E\right] \tag{1}$$

(where B_T is the total toxicant mass in an organism, C_W and C_F are the concentrations of toxicant in the environment and food, F and E are the mass fluxes of food and faeces, and K_W is the unit conductance.)

This equation, based upon thermodynamic potential, represents the chemical exchange that occurs between fish gill membranes and intestinal linings from both the respiration and ingestion of contaminated materials [4]. Metadata for this process description should include any organisms and toxic compounds for which it has been validated, as well as the sensitivity of the parameters to changes when applied to other fauna, such as relevant physiologic or pharmacokinetic differences between aquatic and terrestrial organisms or between vertebrates and invertebrates.

The multimedia science of MIMS is fundamentally different from the traditional means of evaluating environmental chemical exposure risks (for nonhumans or humans for that matter). This difference in treatment relates to the use of environmental concentrations as surrogates for those concentrations internal to an organism. It is intended that biota be treated as media in their own right, since the accounting of contaminant residues in tissues is of interest. As pointed out by Barber [3], while an LC_{50} or LD_{50} toxicological benchmark has been the traditional approach to evaluating exposures in predictive chemical risk assessments, the use of this statistical model has little, if any, relevance to the internal concentrations of chemical residues known to cause physiological effects. Simply stated, bioaccumulation requires explicit treatment with the best available modelling techniques, rather than relying on scenarios of hypothetical environmental concentrations.

This change in perspective is actually an intuitive one, if it is considered that for many contaminants of interest the biological component of a system are a media compartment in their own right. While tracking the fate of methyl mercury, cadmium, or some other persistent bioaccumulating toxic substance it is both more useful and practical to treat biota as a media requiring explicit accounting of mass flow in and out of that control volume. This is analogous to the manner in which the other media are treated, and is important if material cycling is the research topic. Such a utilisation of biota is different than using an organism as an exposure endpoint for chemical uptake. There is sufficient justification to develop multimedia fate and transport approaches that integrate biota as another important system component.

It must be remembered that chemical and other toxic hazards are only a few of a much larger list of relevant environmental stresses on biota. Land-use change, leading to habitat loss and fragmentation, nutrient enrichment, and increases in the rate of sedimentation are important in any comparative risk evaluation. What is needed is a means of evaluating the multiple stresses in concert and not on a case-by-case and stressor-by-stressor basis. Figure 5 shows how the risk assessment and management process could be facilitated by a single software framework, beginning with the creation of the problem-solving environment and continuing through iterations of conceptual model refinement and evaluation of alternative management options. It remains to be seen as to whether this type of automation and computer-aided functions are desired for risk assessors and/or managers.

An additional caveat must be made. A model is an abstraction of the world, of the subject or system or process of interest. Of necessity a model is a simplification. Models are constructed with many assumptions, boundaries, and assumptions about transfers across those boundaries. A model is deemed the best in an instance where it performs its function well in the most efficient way. We are not intending to duplicate nature via software codes. In essence, the more detail that is required involves more computational requirements and can become so much an explicit representation that it mimics nature– and requires just as much time to produce an answer that it is no longer a model at all. Rather, the goal is to produce the most flexible modelling system possible given that one assessment may require a screening level approach and another a more detailed description of the environmental situation, with a high degree of dimensionality and spatially explicit chemical and biological process descriptions.

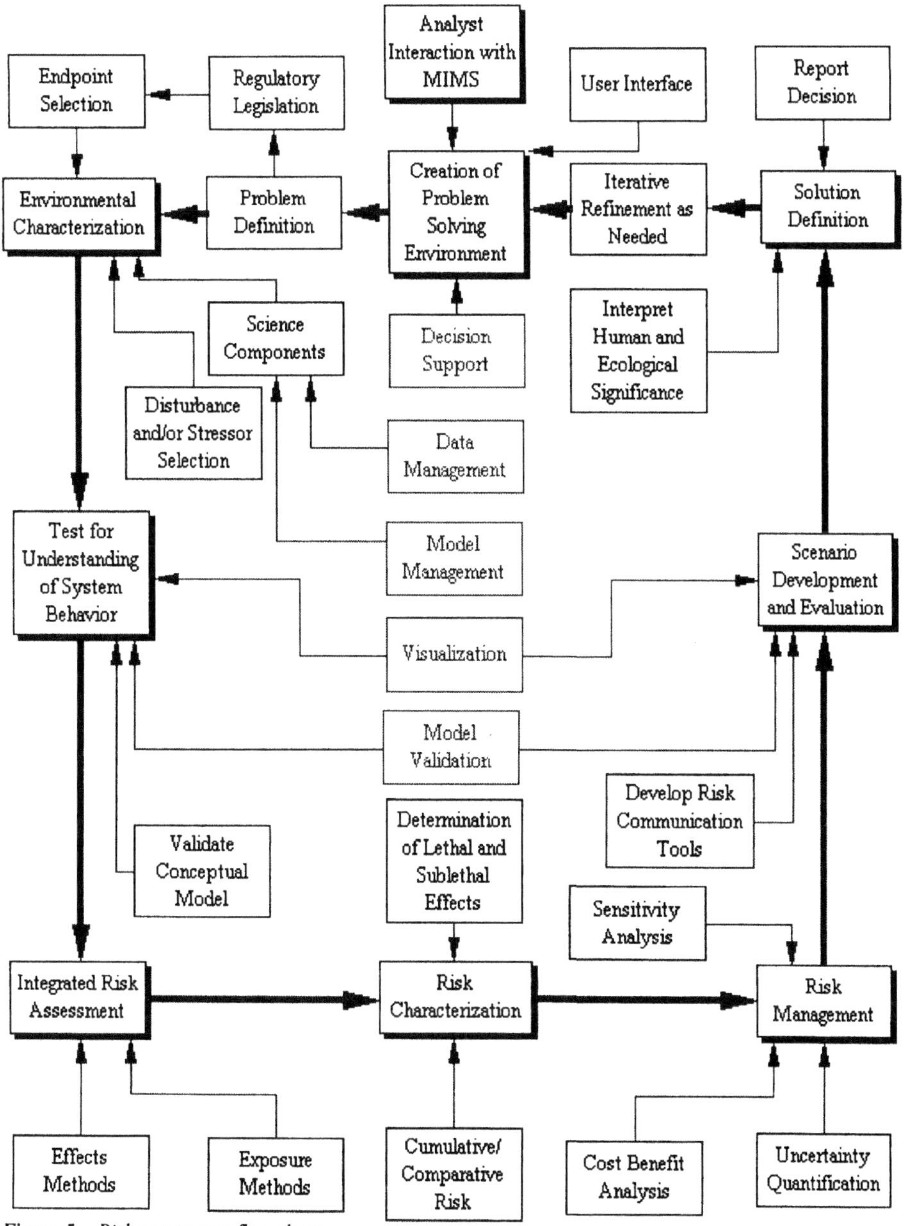

Figure 5 *Risk assessment flow chart.*
This proposed flow chart outlines the risk assessment stages within a software environment. The bold arrow is the risk assessment cycle, the inner components are more detailed supporting activities. The inner row of components are the functions provided by the software framework, enabling and automating many of the time-consuming aspects of a risk assessment / risk management.

4. Conclusions

We should always be able to answer the question: What are we attempting to protect and to what extent? In addition to a clear understanding of this motivation we must also show how we will achieve the research and management goal that we have set. That is, given the range of analysis methods, testing, and model development possible in the solution of complex environmental problems, a consistent, coherent and integrated plan must be articulated that will show in each instance how we intend to assess and manage a given endpoint. The two important areas that sustainability addresses for the management community are 1) being distributed in space to provide information where there are problems and 2) having the necessary information to establish causality.

Risk assessment and risk management have been described as a paradigm [13], a framework [14] and in more casual terms as a process, method, or analysis. Risk assessment (intended to include management, hereafter referred to as assessment) is simply a means to an end that end being a decision for regulating human actions and managing the environment. It is a means of understanding an environmental problem and evaluating the best solutions given a variety of alternatives and their associated costs, likelihood of success and means of measuring that success. Risk assessment can be thought of metaphorically as a blueprint for making decisions, for constructing the types of data and methods of analysis (including statistical or mechanistic models, laboratory and field studies) required in the assessment, as well as how to include the best available scientific information in framing of the management decision for a given environmental situation.

This blueprint is of necessity a general one, as each situation is unique and requires a risk assessment that matches the scientific, environmental, social, and economic contexts of the particular problem. In this manner the blueprint is a general guideline, rather than a step-by-step means of solving environmental problems. Given the diversity of possible contexts that surround a problem, more often than not a solution to one situation will not be applicable to others. Regardless, both the scientific and technical approaches should be flexible enough to test multiple hypotheses. MIMS is particularly useful in this regard, since it is not usually feasible for an investigator to test multiple versions of the same model. An analysis of structural sensitivity, by varying the number of state variables and/or the connections of these components to analyse the appropriateness of the candidate model, is not usually practised in modelling. What is usually developed is the single best guess at system representation. The use of more flexible, object-oriented techniques enables a framework to quickly create multiple, but slightly different, instances of the various science components for such tests.

The results of a sustainability assessment are tractable management endpoints in that the risks are expressed as events per unit time and not probability estimates. Examples of this include the number of extinctions predicted to occur over a 50 year period, the decrease in types of habitat in a region or the reduction in edible sport fish from current levels. These findings clearly communicate the nature of the risk, which is the ultimate goal of any effective risk assessment for decision-making [14].

In practice, predictive risk assessments are useful for regulating specific human activities, while a sustainability assessment is a more useful starting point for environmental protection. There is a useful functional dichotomy between the approaches. What has been conveyed is an appreciation of the scientific challenges to environmental protection, as well as the technical needs that exist for making the science accessible to both the environmental research and management communities. There is a wealth of science that has been essentially locked within legacy codes, and the information as to the formulation and application of many legacy models is not documented completely. Because of this, the use of models for a sustainability assessment for which they were not intended is problematic, and the continued modification and improvement of these models is an involved and challenging process. It may well be that a wealth of scientific information may be lost as new codes will be written to replace them, since the translation of programs into other languages and constructs often requires expert knowledge of the software to begin with. A more open scientific model development process improves not only more efficient and collaborative, it enables a more thorough review of the underlying science itself. Risk assessment and environmental modelling may be many years away from the degree of predictive and practical utility as expressed in a MIMS (Figure 5), but if we are to reach such a goal it will require a careful appraisal of the science and technology that we practice.

5. Notice

This paper has been reviewed in accordance with the U.S. Environmental Protection Agency's peer and administrative review policies and approved for presentation and publication. Mention of trade names or commercial products does not constitute endorsement or recommendation for use.

6. Acknowledgments

Many of the ideas for creating this paper resulted from discussions with my modelling colleagues at the U.S. EPA Ecosystems Research Division, Athens, GA. I am grateful to Ray Lassiter, Craig Barber, Gerry Laniak, Larry Burns, Luis Suarez, Brenda Rashleigh, Bob Baca and others in the Ecosystem Assessment Branch for the constructive comments as we attempted to find our way through these manifold, complex challenges to effective ecosystem protection based on our ecological research. Suggestions from Ray Lassiter improved the final manuscript.

7. References

1. Acock, B. and Reynolds, J.F. (1997). Introduction: modularity in plant models. *Ecol. Model.* 94, 1-6.

2. Barber, M.C. (1994a). Approaches to modelling. Pages 145-148, In: R.J. Kendall and T.E. Lacher, Jr. (Editors), *Wildlife Toxicology and Population Modelling: Integrated Studies of Agroecosystems*. CRC Press, Inc., Boca Raton, 576 pp.

3. Barber, M.C. (1994b). Modelling ecological impact of pesticides on avian populations: Synthesis. Pages 201-204, In: R.J. Kendall and T.E. Lacher, Jr. (Editors), *Wildlife Toxicology and Population Modelling: Integrated Studies of Agroecosystems.* CRC Press, Inc., Boca Raton, 576 pp.

4. Barber, M.C., Suarez, L.A., and Lassiter, R.R. (1988). Kinetic exchange of nonpolar organic pollutants by fish. *Environ. Toxicol. Chem.* 7, 545-558.

5. Benz, J., and Knorrenschild, M. (1997). Call for a common model documentation etiquette*Ecol. Model.* 97, 141-143.

6. Bird, S.L. (1994). Field and exposure modelling in terrestrial ecosystems: A process approach. Pages 149-159, In: R.J. Kendall and T.E. Lacher, Jr. (Editors), *Wildlife Toxicology and Population Modelling: Integrated Studies of Agroecosystems.* CRC Press, Inc., Boca Raton, 576 pp.

7. Burns, L.A., and Cline, D.M. (1985). Exposure Analysis Modelling System (EXAMS): Reference manual for EXAMS II, EPA/600/3-85/038, U.S. EPA, Athens, GA.

8. Burns, L.A. (Editor) (1992). PIRANHA: Pesticide and Industrial Chemical Risk Analysis and Hazard Assessment, Version 2.0, U.S. Environmental Protection Agency, Athens.

9. Carsel, R.F., Smith, C.N., Mulkey, L.A., Dean, J.D., and Jowise, P. (1984). User's manual for the Pesticide Root Zone Model (PRZM), EPA/600/3-84/109, U.S. EPA, Athens, GA.

10. Hoch, R., Gabele, T., and Benz, J. (1998). Towards a standard for documentation of mathematical models in ecology. *Ecol. Model.* 113, 3-12.

11. Johnston, J.M., Novak, J.H., and Kraemer, S.R. (1999). Multimedia integrated modelling for environmental protection: Introduction to a collaborative framework. *Env. Mon. Assess.* In press.

12. Science Advisory Board. (1999). Integrated Risk Project Report.

13. Suter, G. W., II (Ed.). (1993). Ecological risk assessment. Lewis Publishers, Boca Raton.

14. The Presidential/Congressional Commission on Risk Assessment and Risk Management. (1997). Risk Assessment and Risk Management in Regulatory Decision-Making. Final Report Vol. 2. (http://www.riskworld.com).

15. Williams, J.R., Jones, C.A., and Dyke, P.T. (1983). A model for assessing the effects of erosion on soil productivity. In: W.K. Lauenroth, G.V. Skogerbee and M. Flug (Editors), *Analysis of ecological systems: State-of-the-art in ecological modelling.* Elsevier, New York.

COMPARING TWO ALTERNATIVE POLLUTANT DISPERSION MODELS AND ACTUAL DATA WITHIN AN ENVIRONMENTAL HEALTH INFORMATION PROCESSING SYSTEM (EHIPS)

BORIS BALTER, M. STAL'NAYA, VICTOR EGOROV
Space Research Institute, Russian Academy of Sciences
Profsoyuznaya 84/32,
Moscow, 117810,
Russia

Abstract

This paper presents the test results for an air pollutant dispersion modelling unit working within a larger software system EHIPS designed for environmental health information processing. We start with the description of the whole system and then proceed to the modelling unit, which uses two standard dispersion models - one Russian (OND-86) and one American (ISC3ST). The peculiarity of this unit is just in its links to other components of the system, which opens the way to multifold usage of the modelling results, including easy comparison between the model and the data obtained from actual measurements.

We compare the two models and actual measurements on the basis of a one-year data set obtained in an industrial city (Cherepovets, RF). The overall agreement between all three may be characterised as satisfactory. Our interest lies primarily in analysing the dependence of discrepancies between the models and between models and data as a function of pollutant, time, etc. We also discuss the influence of expected deficiencies in pollutant source data on accuracy of modelling. We try to draw lessons for interpreting the risk assessment figures obtained through modelling.

1. What is EHIPS

In a narrow sense, EHIPS is the software that processes data and model calculations related to the chemical pollution of environment and population health status.

The main features are:

- It outputs the 'portraits' and forecasts of the health hazards associated to environmental pollution;
- assesses the relative severity of diverse hazards by the user-set criteria;
- selects priorities for pollution control;

J.B.H.J. Linders (ed.), Modelling of Environmental Chemical Exposure and Risk, 151–164.

- estimates the certainty or uncertainty of inferences made from existing data;
- issues requests for collection of additional data, if necessary.

In perspective, EHIPS will also:

- calculate economic damages associated to the health loss due to environmental pollution;
- select options of pollution control that have justifiable costs;
- optimise the control schedule using the cost-benefit analysis;
- track the actual effects of an adopted pollution control plan and issue corrections to it, if necessary.

The primary users of EHIPS are the official organisations entitled to control of environmental status and related health effects: SanEpid, Ecological committee and municipal administration. The primary uses of EHIPS for them are:

- identifying the priority problems;
- distributing the responsibility for them between 'guilty' parties and, if necessary, support sanctions in court;
- evaluating the potential health hazard of construction and other projects;
- optimising the local environmental monitoring system to enhance the value of information;
- forming the public opinion on urgency of diverse environmental issues and balancing it with economic realities;
- optimising actions to alleviate environmental health problems and tracking the results;
- detecting the new hazards that call for immediate reaction.

2. What Problems Addresses EHIPS?

There exists a lot of software for environmental problem analysis: pollutant dispersion modelling, health risk calculation, environmental epidemiological studies etc. However, each software addresses just a specific aspect of data processing. In contrast, EHIPS was designed to connect these multiple aspects and to make one way of data analysis build on another. This drive toward universality can help in the following problems.

Usually, the software made for regulatory purposes (e.g., compliance check) uses some fixed algorithm to calculate concentrations and risks, fill data gaps and so on. However, environmental issues are scientifically intricate and usually need much experimenting and data fitting to obtain a defensible result. Such scientific analysis is supported by completely different software, if at all. EHIPS synthesises regulatory calculation thread and scientific analysis thread. The latter set parameters for the former and, in its turn, checks the results and, if necessary, adapts parameters.

EHIPS does not call for creation of special monitoring systems. It is oriented towards standard databases that exist in Russia on pollutant emissions, measured concentrations,

morbidity, mortality etc. Such data mostly lie dead or are used very superficially because before they can be effectively used, they need a lot of preparatory checking, linking to other data, fitting into a specific problem framework. EHIPS brings data deposits back to life by providing a flexible interface (checked on several dozen types of Russian databases) through which data are imported into the universal processing engine and thus are made active.
All types of data mentioned above have a common set of basic operations:

- hierarchical aggregation;
- 'section' choice and review;
- normalisation;
- hotspot/outlier detection etc.

Universal statistical analysis packages support such operations for any type of data, but they contain no specific environmental 'machinery'. In contrast, environmentally oriented packages are normally confined to specific types of data, so that, e.g., mapping operations are not readily extensible from concentration contours to morbidity statistics. EHIPS standardises handling of diverse data types by embedding each in a 'dataspace' with the following 'axes':

- time,
- territory,
- pollutant,
- population group,
- diagnosis.

Each axis is hierarchically organised. All charting, mapping, statistics, etc. is independent on dataspace choice. Thus, EHIPS is potentially a standardisation tool for environmental analyses.

Normally, environment-oriented software uses either results of direct measurements or results of computer simulation, but not both. However, none of the two is beyond doubt, and it is advisable to use them concomitantly, for mutual check and adjustment. By supporting this regime, EHIPS provides for verifiability of results.

3. EHIPS Development Status

EHIPS was initiated in 1995 by the environmental modelling group in the Space Research Institute, Russian Acad. Sci., Moscow. In 1996 - 1997 the work was continued by the same group and some additional staff in the framework of RF Environmental Management Project (EMP), and in 1997 - 1999 again by the Space Research Institute. Three configurations of EHIPS have emerged:

- The **minimal one** runs on a single autonomous PC where all databases reside. It includes a restricted set of data modules and functions (none of those in small font) and is normally used by a single expert in a municipal/regional state agency. It is oriented mainly toward air pollution.
- The **intermediate one** runs on a local network of several PCs with databases arbitrarily distributed over the network. It has the full set of data modules and

functions and is supported by a special organisational structure - environmental health data processing centre - within the same state agency framework.
- The **maximal** - networked - configuration consists of several systems of the intermediate type installed in different regions and linked through Internet to a centre in Moscow that provides the methodological support, reference databases, materials, etc. to the entire network.

The first configuration is basically ready for installation, the third one exists only in a design document, and for the second one, some work has been done and some not. Presently, the federal SanEpid officials consider EHIPS as an option for a standard nation-wide tool for environmental health data analysis within local SanEpid services. Before that, EHIPS yet has to pass the necessary testing and certification stages.

4. Functions

The main functions of EHIPS are as follows.
- **Data overview** generates tables, charts, maps and histograms of data unfolding along one or several dataspace axes mentioned above. Data can be handled as absolute values or relative to some data subset taken as a reference frame. Unfoldings along similar axes can be compared; in this way, model calculations are gauged against actual measurements. This function, although routine, is critical for an effective interface to end users.
- **Data refinement** is the process of selecting the right data subset and aggregation scheme for the problem at hand. It includes stating the problem at several hierarchical levels at once; detecting outliers; discriminating them from meaningful hotspots; selecting the data subset with tightest connection between environmental and health indicators, etc. This is a prelude to data analysis, and it can be repeated afterwards when necessary.
- **Hotspot search and verification** is one of the simplest functions, but the one most called-for by the end users. Hotspots are contingent intervals of time, compact territories or portions of any other dataspace axis where the values of some data type (concentration, mortality etc.) do significantly stand out from the background. 'Values' can also mean some model quality indicator, e.g., residual dispersion, so that the hotspot is where the model performs best. Hotspots are used either for further in-depth investigation or as an end product for decision-makers. An important part of this function is determining the hotspot stability and excluding the artefacts of data processing.
- Further analysis leads from hotspots to the **ranked set of priority hazards**. Roughly speaking, the set of priorities is the set of hotspots multiplied by the user-set matrix of importance. A priority can combine diverse types of data related to the same hazard, so that priority ranking should use multicriterial analysis. The same is true when there are several users with different criteria of importance. In the risk paradigm, priorities are where the risk assessment ends and the risk management begins. Within the risk management data processing, priority hazards are further multiplied by the costs of control, and the cost-benefit analysis is performed.

In contrast to the functions above, the **model construction and model parameter setting** belongs to the research thread of EHIPS. It includes empirical models, like regression, physically based models, e.g. pollution dispersion, and models that embody the expert judgement, e.g. risk formation and risk expression in morbidity. All these types are treated in a unified manner: model predictions are compared to actual data and the model parameters are obtained from the best fit. This can be done automatically or by experts. The models are used to generate 'simulated' data, which are then processed into hotspots, priorities, etc. Some results of model fit (e.g., pairs of tightly correlated environmental and health indicators) are a direct output for decision-makers, since it helps to identify hazards.

Uncertainty estimation is an ancillary function, which accompanies all stages of data processing. It combines all sources of uncertainty: statistical variation, model inaccuracy, data errors, etc. It finally produces the quality index for EHIPS output information.

5. Source Data

The following databases should exist for the territory of interest and for at least 1 year.

- Emission sources and intensities;
- pollutant concentrations measured several times a day at several locations; standard daily meteorological data;
- GIS basis with population data and environmentally significant objects;
- personal registration of admissions to selected medical institutions and/or of ambulance visits;
- personal mortality registration.

In the **minimal configuration** of EHIPS, some of these databases may lack and be substituted by model calculations.

In the **maximal configuration**, the following additional databases should be present:

- demographic parameters of the population for at least 10 past years; parameters of the social-economic status of the population;
- costs associated to each aggregated type of health damage;
- technologies implemented at pollution sources, their optional substitutes and associated costs.

The data sources of EHIPS are shown in Figure 1.

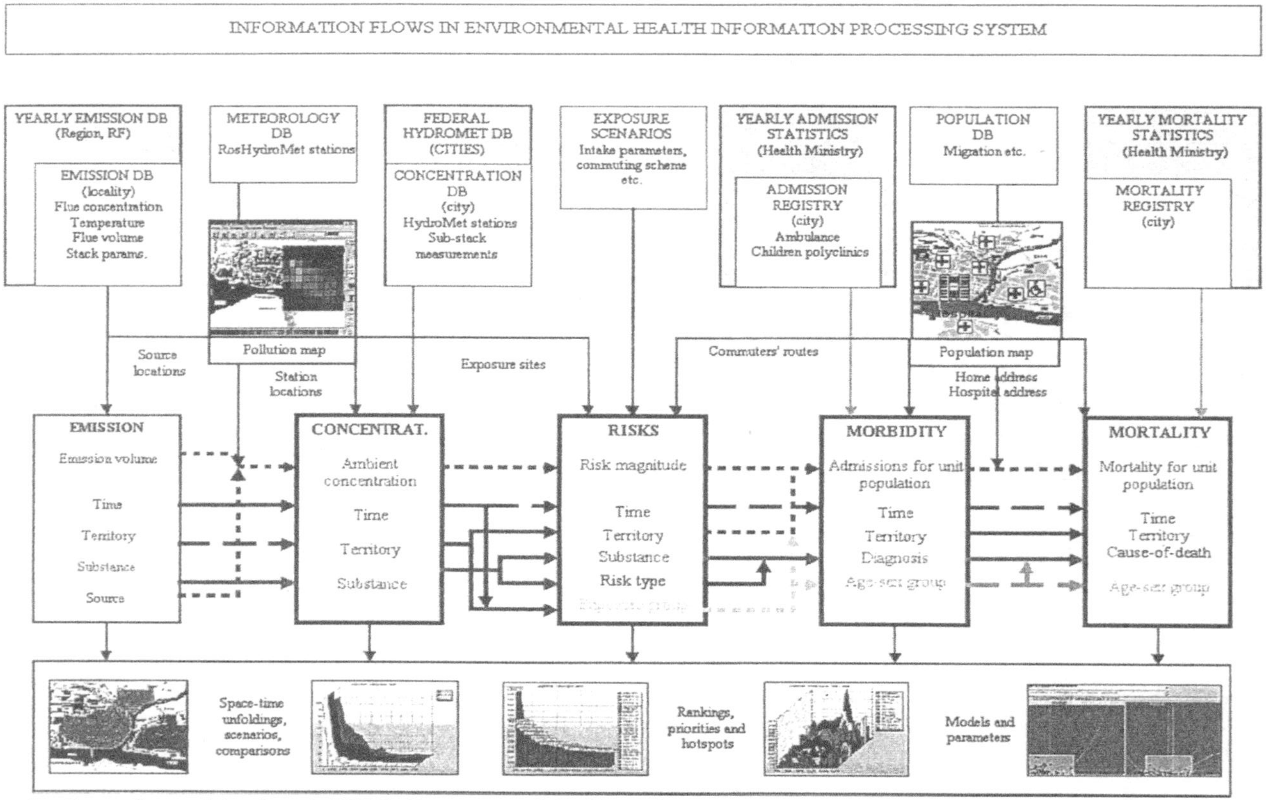

Figure 1 ***Source data and dataflows in EHIPS. Gray squares: federal-level data; white squares within: regional-level data. Rectangles outlined in bold are main processing units each specialised to a data type and forming an information processing pipeline. Within each unit variables are listed across which data are unfolded in the unit. Below: forms of processing output.***

6. EHIPS Structure

EHIPS consists of sequentially linked modules, one for each data type. Presently, there are 5 modules for emissions, concentrations, risks, morbidity and mortality. Extended EHIPS structure will include also modules for integral indices of health loss, related economic loss, cost/benefit criteria, choice of control measures, control implementation plan, and technological processes that lead to pollutant emission.

The background data processing consists in exchange of information between modules. Normally the data is passed along the module sequence, transition between modules being performed by models (e.g., from emissions to concentrations by the pollutant dispersion model; from concentrations to risk by the exposure model, etc.). In a module, the model calculations are harmonised with the measurements taken from databases linked to the module, and then passed on. In extended EHIPS, this process includes the choice of emission control measures, after which the calculation starts with new emissions. Thus, the cycle is closed and runs in circle until a steady state is reached so that a plan of control and respective evolution of hazard are obtained. On this background, EHIPS builds up its functions described above. They are performed by the functional blocks that work with all data modules on equal footing. There is the data overview block that performs tabulating, charting, mapping; the statistical block that performs correlation, regression, cluster and pattern analyses and so on.

Finally, EHIPS includes the service modules that perform links to databases, form the output document, support the network data exchange, etc.

The backbone structure of EHIPS in minimal configuration is shown in Figure 1. The information flow in the maximal configuration of EHIPS is shown in Figure 2.

7. Dispersion Modelling Unit in EHIPS

Dispersion modelling in the minimal version of EHIPS includes only propagation of flue in air (see upper left part of Figure 1). So, it is just a tiny part of overall data flow. Still, it is important because there is a well-studied physical process behind it. So, it should serve as a template for modelling and model validation.

We use two concurrent models of flue dispersion: ISC3ST developed by US EPA and OND-86 developed by Russian Voeikov Geophysical Observatory. Both are used for regulatory purposes in respective countries. There are many minor differences in scientific approach between these models; the major difference is that ISC uses Gaussian approximation of plume profile from the start, while OND introduces it postfactum, after reducing differential equations of propagation to algebraic form.

The general structure of both algorithms is the same, and it is shown in Figure 3.

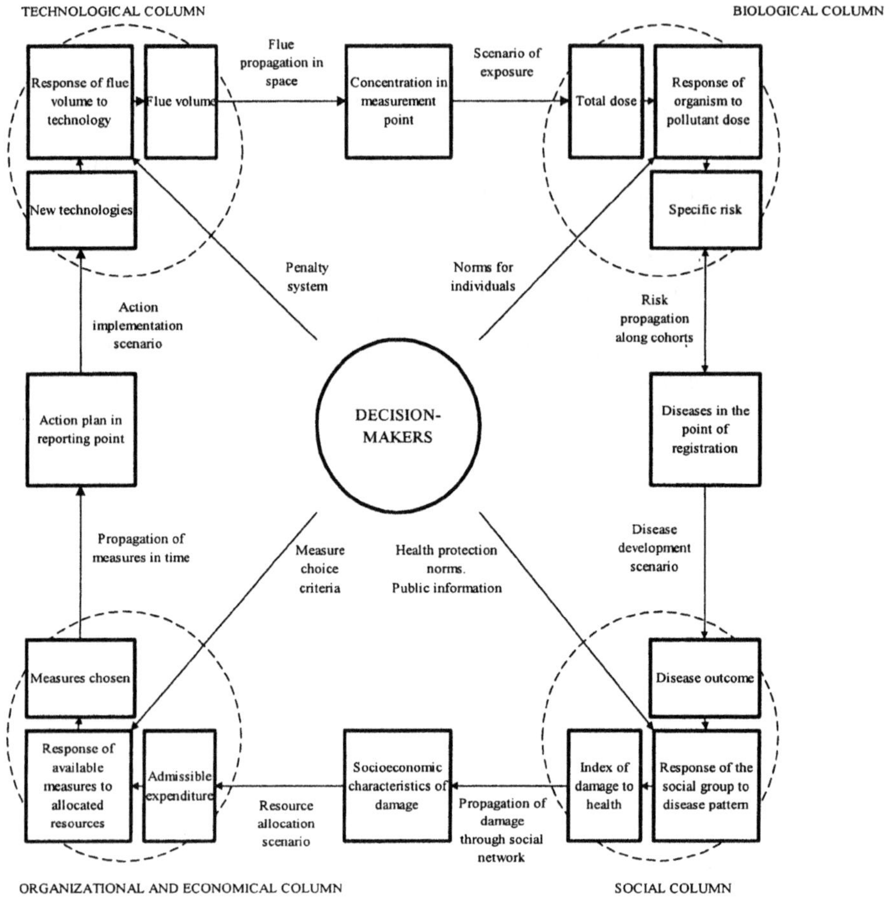

Figure 2. Information processing flow in EHIPS.

The major practical issue is that OND is targeted at calculation of absolute concentration maxima at the worst meteorological conditions, while ISC can predict concentrations for any current meteorological situation and is therefore more fit for short-time monitoring and forecast. However, this limitation is not inherent in the OND algorithm, and we extended it using the original description of OND equations so that the two models can be compared at the equal footing. Still, there remains a degree of freedom in transferring the atmospheric stability index between the two models, and this can sometimes change the result by as much as the factor of 2 or 3. Therefore, rather than sticking to absolute values, we paid most attention to the behaviour of model predictions across time, space, and other 'unfolding variables'.

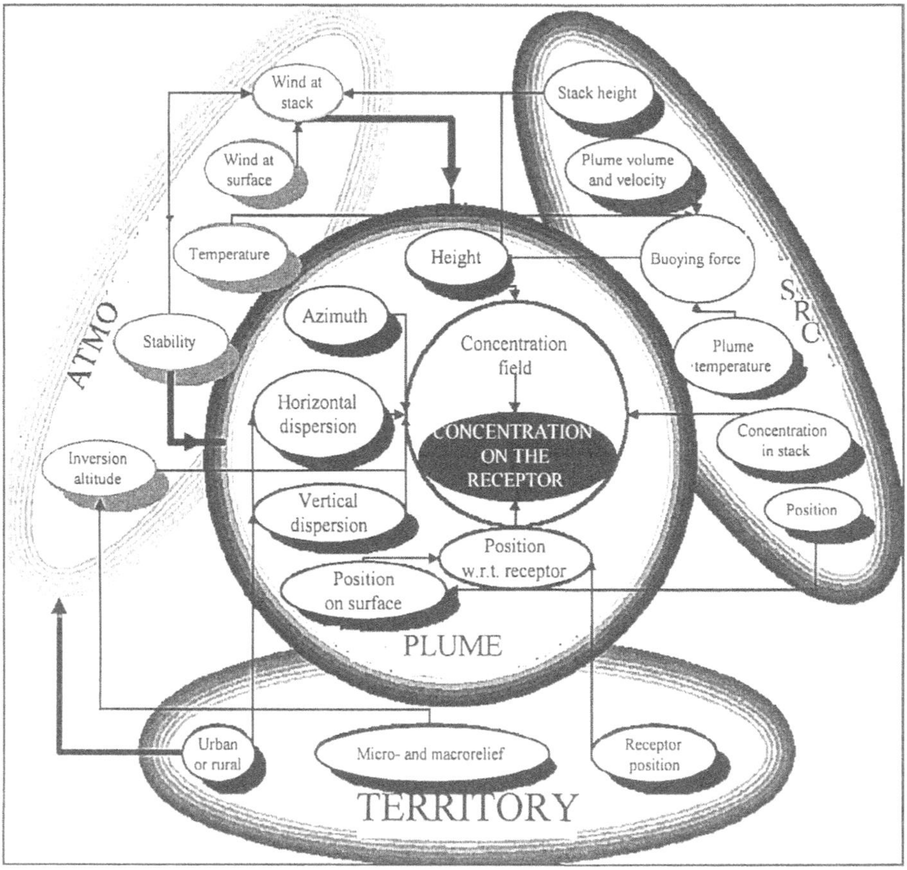

Figure 3. Parameterisation and information flow structures common for ISC and OND. Arrow thickness indicates the relative weight of dependence between parameters it connects.

8. Results of Modelling

In the following figures we present an example of colour-coded concentrations calculated on a grid of square cells that cover the residential area of Cherepovets City. Colour codes can be read off the scale above the map. The small icons to the upper left of the grid represent the flue emission sources within the largest plant in the city: the Severstal steel plant. However, there exist also other emission sources which can contribute significantly (at least for certain locations and certain pollutants). In addition, flue emission rate was taken from yearly reports rather than from actual measurements on the stack. All this amounts to a considerable uncertainty about emission sources. Therefore, model comparisons to measured concentrations should be 'handled with care'.

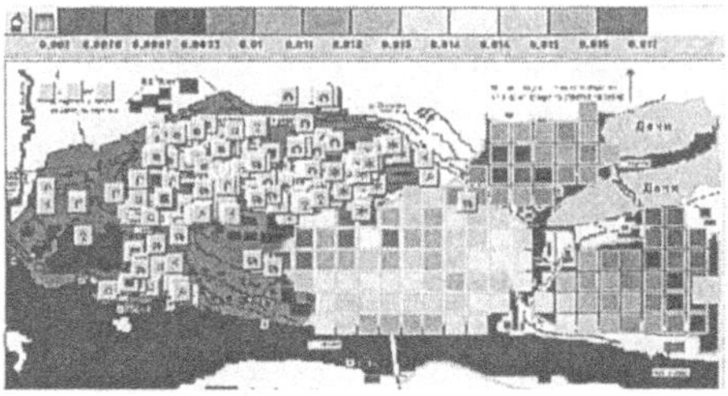

Figure 4. ISC3ST: yearly averages of four-times-a-day NO_2 concentrations calculated from current meteo.

Figure 5. OND-86: yearly averages of four-times-a-day NO_2 concentrations calculated from current meteo.

We see that the absolute values consistently differ by approximately a factor of 2, but there is an obvious similarity in the spatial pattern. Its most important feature is a hotspot aside from the flue 'mainstream' - in the northern residential area (more pronounced in OND). Its presence was important for regulatory purposes and also correlated with the higher concentrations measured in this area.

9. Comparisons

For comparisons between models or models and data we use the tabular/graphics units of EHIPS to which we feed the concentrations. A set of concentrations (say, from one model) is chosen as a reference frame, and another similar set (from another model or from actual data) is plotted with respect to this frame across time, or some other unfolding variable, or turned into a histogram. As a result, we obtain an overview of discrepancies between models or between a model and actual data. Some examples are shown in

Figure 6 to Figure 11. (We also introduced some filtering to remove suspected outliers). In these calculations, we used ISC3ST. The range of ratio values is approximately 0.1 - 10. The figures are captured directly from the program's screen so inscriptions are in Russian. See figure captions for explanations.

For most pollutants, model concentrations are considerably lower than measured. That is partly attributable to an incomplete list of pollutant sources. We can make the reasons for discrepancy more clear by unfolding the model/data ratio across time since different sources behave differently in this respect.

For NO_2, in summer months, the model is almost exact (ratio close to 1) and in winter it considerably underestimates concentrations. This is probably because thermal power plants used for heating, which were unaccounted for in the model, are not operational in summer, and in winter they are a major source of NO_2.

For CO, the models underestimate of concentrations has no obvious seasonal dynamics. This is probably because a large part of CO is due to transport (unaccounted for in the model), which has no pronounced seasonal dependence.

For dust, the model, on the average, was almost exact. That was expected because there were no major sources of dust outside the industry. Interestingly, there were events when measured concentrations were less than modelled. One possible explanation for that lies in the variation of emission intensity (temporary decreases).
As an example of data viewing options, we give the same ratio averaged over a whole city district, now not as a histogram, but as a ranking. In this way, we can quickly identify the largest discrepancy events, to study them in detail.

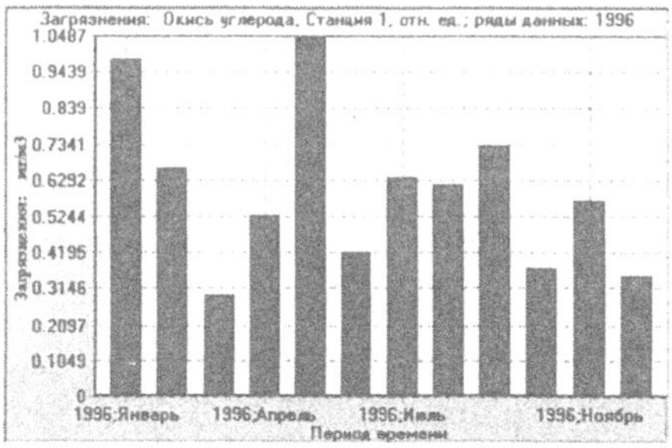

Figure 6. Ratio of modelled CO concentration to actual data (X-axis). Histogram of 4-times-a-day values over 1 year (Y-axis: occurrence per year). Monitoring station 1, year 1996.

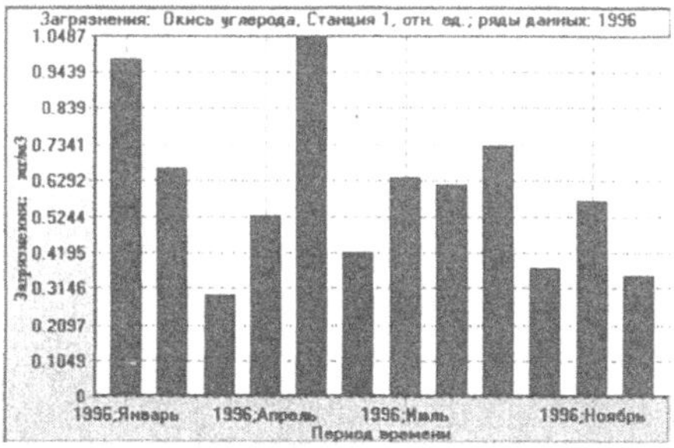

Figure 7. Monthly time series for the same ratio (plotted as Y-axis). X-axis: 1996, January to 1996, December. Monitoring station 1, year 1996. The filtering used was the same as in the histogram.

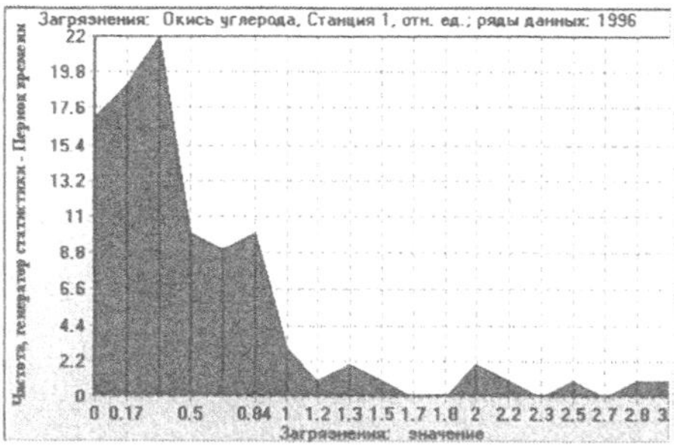

Figure 8. Ratio of modelled NO_2 concentration to actual data (X-axis). Histogram of 4-times-a-day values over 1 year (Y-axis: occurrence per year). Monitoring station 1, year 1996.

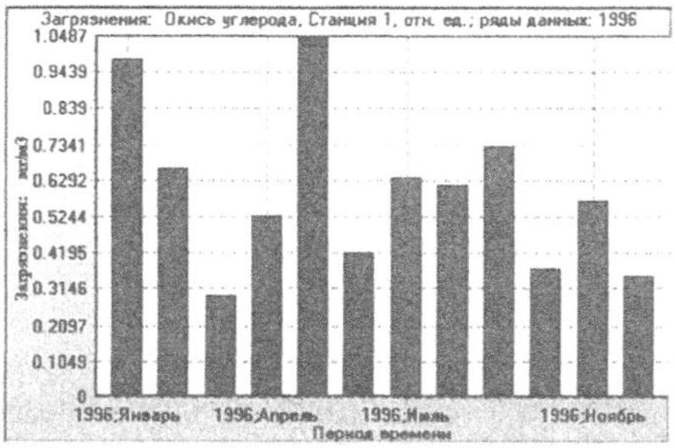

Figure 9. Monthly time series for the same ratio (plotted as Y-axis). X-axis: 1996, January to 1996, December. Monitoring station 1, year 1996.The filtering used was the same as in the histogram.

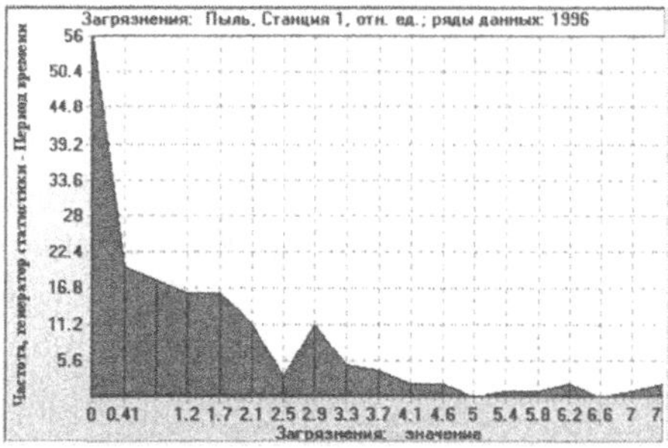

Figure 10. Ratio of modelled dust concentration to actual data for a single monitoring station (X-axis). Histogram of 4-times-a-day values over 1 year (Y-axis: occurrence per year). Monitoring station 1, year 1996.

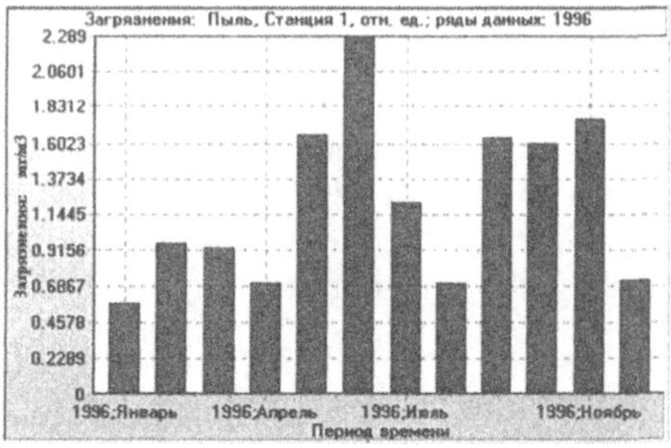

Figure 11. Monthly time series for the same ratio (plotted as Y-axis). X-axis: 1996, January to 1996, December. Monitoring station 1, year 1996. The filtering used was the same as in the histogram.

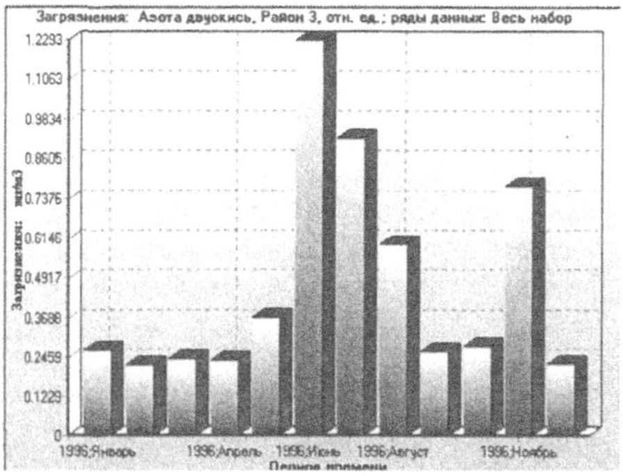

Figure 12. The ratio of NO_2 model concentrations to measured data averaged over a city district (plotted as Y-axis). X-axis: 1996, January to 1996, December. City district 3, year 1996. Monthly averages of four-times-a-day ratios.

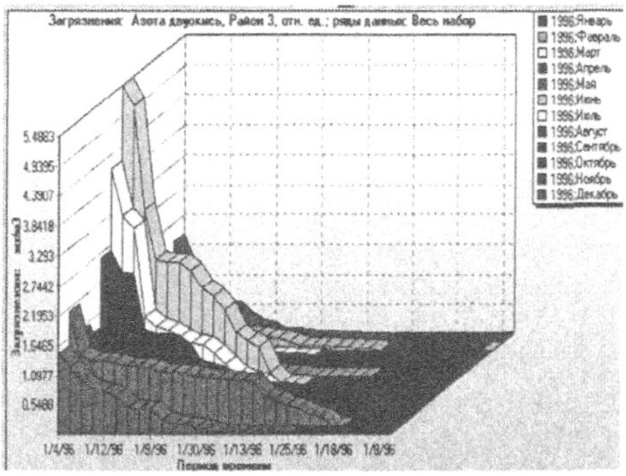

Figure 13. The ratio of NO_2 model concentrations to measured data (plotted as Y-axis) averaged over city district 3 (monthly ranking of four-times-a-day ratios). X-axis: days of the respective month ranked highest concentration ratio first.

Another direction of study was the comparison between ISC and OND. For NO_2, the models give comparable results 'on the average' but discrepancies by a factor of 2 are not unusual. That is primarily due to sensitivity of results to modelling details when the cell, for which modelling is performed, is on the plume's tail. Once more, as seen in *Figure 5*, the spatial patterns are very much the same for both models.

10. Using the Results

The primary use of model concentrations in EHIPS is for health risk assessment. We are primarily interested in non-cancer risks for which standard risk assessment schemes do not provide a good estimate. EHIPS uses for that a special risk model developed by Prof. S.Novikov. It combines inputs from all calculated non-carcinogeneous pollutants.

The resulting risk map is shown in Figure 14. It is calibrated so that zero risk corresponds, roughly speaking, to non-observable effect level, and the risk equal to 1, to death outcome. As shown in the histogram of daily risks (Figure 15), the prevalent input is formed by the dust, which is known to be related to total mortality.

Note that, due to overlapping of spatial patterns for several pollutants, the risk pattern is slightly different from those shown in

Figure 5 etc. The northern hotspot took the form of an elongated 'concentration ridge'.

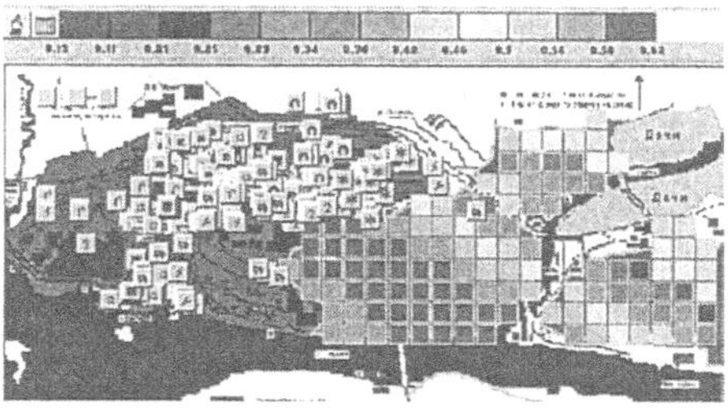

Figure 14. Noncarcinogeneous risk displayed on the city plan: 1-year average of daily values calculated from ISC3ST model. Top: color scale for relative risk index. Colored squares: cells of calculation grid. Square icons: emission sources, icon showing the type. Black squares: industrial buildings.

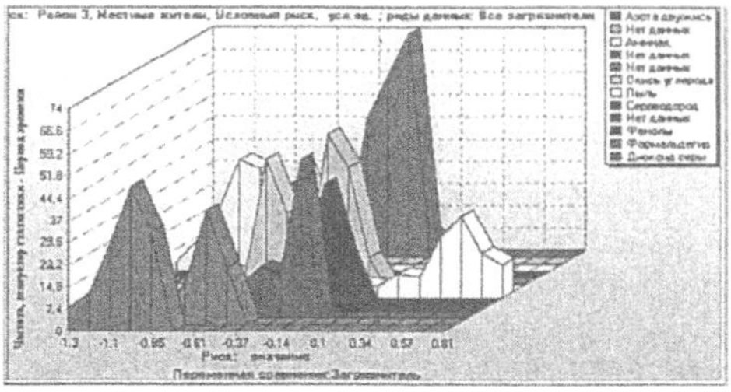

Figure 15. Noncarcinogeneous risk: 1-year histograms of daily risk values (plotted as X-axis) calculated from the ISC3ST model for major measured pollutants. The list of pollutant color codes (top right): NO_2; no data; NH_3; no data; no data; CO; dust; H_2S; no data; phenols; formaldehyde; SO_2. Yearly occurrence of daily risk values plotted as Y-axis.

11. Discussion

We have drawn the following preliminary conclusions from our modelling experiments in EHIPS.

1. OND and ISC give compatible yearly averages but can diverge by the factor of 2 to 3 in one-shot estimates. They can be used concurrently for mutual check.
2. Even if models underestimate actual concentrations considerably, they can give reproducible (and hence, hopefully, correct) spatial and temporal patterns. Therefore, they can be corrected to actual measurements so as to 'interpolate' between sparse measurement points.
3. There can be stable hotspots in unexpected locations, and model-based risks in adjacent modelling cells may differ by a factor of 2.

12. Contacts

For additional information and documentation please contact EHIPS team through:
Balter Boris Mikhailovich: Balter@iki.rssi.ru (095)333-4467
Egorov Victor Valentinovich: Egorov@iki.rssi.ru (095)333-3589

REFERENCES

View the **EHIPS Website** (mostly in Russian) at
http://www.iki.rssi.ru/ehips/welcome.htm .

RBCA TOOLKIT: COMPREHENSIVE RISK-BASED MODELLING SYSTEM FOR SOIL AND GROUNDWATER CLEAN-UP

JOHN A. CONNOR, P. E., RICHARD L. BOWERS AND
THOMAS E. MCHUGH
Groundwater Services, Inc.
Houston
Texas
USA

Abstract

The ASTM "Standard Provisional Guide to Risk-Based Corrective Action" [2] expands upon the earlier RBCA standard for petroleum releases to establish a comprehensive process for management of both human health and ecological risks associated with chemical releases to soil or groundwater. Modelling tools can prove helpful at various points of the RBCA site evaluation process, including exposure pathway screening, tiered evaluation of cleanup standards, remedy selection, and design of the compliance monitoring program. This paper provides practical guidelines for the use of modelling tools in the risk-based decision-making process and reviews the *RBCA Tool Kit for Chemical Releases,* a comprehensive modelling and risk characterisation software designed for completion of RBCA Tiers 1 and 2. Key topics include application of modelling tools to risk-based site evaluations; model selection guidelines; applicable analytical models for soil, groundwater, surface water, and air exposure pathways; and use of transient and probabilistic modelling methods.

1. History of Risk-Based Site Evaluation

Historically, chemical releases to soil or groundwater have been managed by requiring restoration of the site to original (background) conditions or to generic regulatory remediation standards often based on laboratory analytical detection limits. With experience, however, it has become increasingly apparent that applying such standards without consideration of the extent of actual or potential human and ecological exposure is an inefficient means of providing protection against the risks associated with chemical releases. In addition, no current technology is capable of restoring many sites to background conditions or reducing chemical concentrations to analytical detection limits. These prob-

J.B.H.J. Linders (ed.), Modelling of Environmental Chemical Exposure and Risk, 165–181.

lems have become more serious as the number of reported chemical release sites has increased. More than 370,000 releases have been reported in the United States from underground storage tanks alone (mostly gasoline stations).

Risk-based corrective action is a process to determine the extent and urgency of corrective action required at a chemical release site based on relative risks to human health and the environment. Depending on known or anticipated risks to human health and the environment, appropriate action may include site closure, monitoring and data collection, active or passive remediation, contaminant, or institutional controls. In all cases, the objective is the same, i.e., to ensure that adequate protection of human health and the environment is provided. The availability of options such as allowing contamination to remain in place or using institutional controls to prevent exposure will depend on applicable national and local laws and regulations.

In 1994, a detailed framework for taking risk factors into account in making corrective action decisions was issued by the American Society of Testing and Materials (ASTM). This guide, finalised in 1995 as the "Standard Guide for Risk-Based Corrective Action Applied at Petroleum Release Sites" [1], was developed in the United States by a committee of industry representatives, government regulators, and environmental consultants. The ASTM "Standard Provisional Guide for Risk-Based Corrective Action" [2], issued in 1998, expands upon the earlier RBCA standard for petroleum releases to establish a comprehensive process for management of both human health and ecological risks associated with chemical releases to soil or groundwater. Currently, 41 state environmental regulatory agencies in the United States have implemented or are developing programs for risk-based corrective action. The majority of these programs use the ASTM framework as the basis for development of a risk-based process tailored to applicable state and local laws and regulatory practices.

2. Overview of Risk-Based Corrective Action Process

Effective risk management at chemical release sites entails i) identification of applicable risk factors on a site-specific basis and ii) development and implementation of appropriate protective measures in the timeframe necessary to prevent unsafe conditions. Key elements of the risk-based site evaluation process include:

- *Exposure Pathway Screening*: Identify potential mechanisms for exposure of human or ecological receptors on a site-specific basis.
- *Risk-Based Cleanup Objectives*: For each complete exposure pathway, evaluate potential for exposure in excess of safe limits based on tiered evaluation of soil and groundwater cleanup limits.
- *Remedy Selection*: Develop risk-based exposure control strategy based on the nature and timing of the potential impact.
- *Compliance Monitoring*: If needed, conduct final compliance monitoring period to confirm satisfactory remedy completion prior to formal case closure.

Further discussion of these process steps and relevant risk-based modelling tools is provided below.

2.1 EXPOSURE PATHWAY SCREENING

The risk-based evaluation addresses the potential for constituent transport from the affected media source zone to a point of contact with a human or ecological receptor via various *exposure pathways*. For most remediation sites, the primary exposure pathways of human health concern are i) groundwater ingestion, ii) soil-to-groundwater release, and iii) soil ingestion, vapour inhalation, or dermal contact. Additional exposure pathways may apply based on site conditions and land use (e.g., surface water impacts, ecological exposures). To pose a risk, three components of each exposure pathway must be present: an affected source medium, a mechanism for constituent transport, and a receptor. In practical terms, exposure pathways may therefore be screened from further consideration based on the presence and mobility of the constituents of concern and the proximity of receptors to the source zone. For example, if an affected groundwater plume is in a stable or diminishing condition, no potential exists for impacts on water supply wells located outside the current plume area.

Pathways determined to be potentially complete should be retained for site-specific evaluation. However, if the preliminary screening analysis shows no complete exposure pathways, no further evaluation is required.

Applicable Data Evaluation Tools: General guidelines for pathway screening are presented in ASTM Standard E 1943-98 [4]. The ASTM "Standard Guide for Remediation of Groundwater by Natural Attenuation at Petroleum Release Sites" outlines practical data evaluation methods for analysis of groundwater plume stability, including historical data plots, estimation of bulk attenuation rates, and modelling methods. The *Natural Attenuation Tool Kit* currently under development for use with the ASTM RNA Standard is described by Nevin *et al.* [5].

2.2 RISK-BASED CLEANUP OBJECTIVES

The RBCA process employs a tiered approach to derivation of risk-based soil and groundwater cleanup goals, with each tier serving to refine the risk analysis based upon additional site data and more sophisticated fate and transport modelling methods. For example, Tiers 1 and 2 of the site evaluation process are amenable to use of simple analytical models to estimate risk-based concentration limits, while more complex and costly numerical modelling methods are reserved for Tier 3 evaluations. In each case, risk-based concentration limits are derived for relevant exposure pathways, receptors, and constituents of concern (COCs) and compared to measured source media concentrations. Source media exceeding these target levels will require either further investigation or remedial action in the timeframe necessary to control exposure. Summary information regarding principal calculation steps is provided below.

2.2.1 Site-Specific Target Level

For a given exposure pathway and COC, the Site-Specific Target Level (SSTL) represents a concentration in the affected source medium (soil or groundwater) that is protective of a human or ecological receptor located at a relevant point of exposure (POE). For example, for the human health soil-to-air exposure pathway, the SSTL is the mean concentration in the affected surface soil zone that will prevent unsafe human exposures via soil vapour or particulate release to air. For each complete exposure pathway, SSTL values for the source medium can be back-calculated from safe exposure levels at the POE using the following general expression:

$$SSTL = RBEL \cdot NAF \qquad (1)$$

Where RBEL = Risk-based exposure limit at POE

NAF = Natural attenuation factor defining natural reduction in constituent concentrations during transport from source to POE.

SSTL values must be developed for each complete exposure pathway and COC. For exposure pathways with multiple POEs (e.g., ambient vapour inhalation by on-site worker and by off-site resident), separate SSTLs must be developed for each POE using the appropriate RBEL value. In general, the RBEL value does not vary among Tiers 1, 2, and 3. Rather, the SSTL value is refined among each successive tier by improving the NAF value estimation based upon more complete site information and more sophisticated data evaluation methods. Determination of applicable RBEL and NAF values is addressed below.

2.2.2 Risk-Based Exposure Limits

The RBEL represents the constituent concentration that is safe for consumption by the receptor, based on applicable regulatory criteria and/or target risk limits. The RBEL applies at the POE, i.e., the likely point of constituent intake or contact by a human or ecological receptor. For each complete exposure pathway and COC, the applicable RBEL must be matched to each relevant POE based on the type of exposure medium (air, water, soil) and the type of receptor (resident, commercial/industrial worker, etc.). For certain exposure media, human health-based exposure limits are specified under applicable regulations, such as Maximum Contaminant Levels (MCLs) for drinking water ingestion or Permissible Exposure Limits (PELs) for industrial air exposure. In the absence of such standards, human health RBELs can be derived for each constituent and exposure medium (air, water, soil) using the following general expressions:

carcinogens:
$$RBEL = TR/(E \cdot SF) \tag{2}$$

non-carcinogens:
$$RBEL = (THQ \cdot RfD)/E \tag{3}$$

where

E	=	effective exposure rate for specified pathway, based on applicable exposure factors (i.e., daily intake rate in mg/day per kg body weight)
TR	=	target risk limit for carcinogenic effects of individual constituents (dimensionless)
SF	=	slope factor for carcinogenic effects of COC $(mg/kg\text{-}day)^{-1}$
THQ	=	target hazard quotient for non-carcinogenic effects of individual constituents (dimensionless) and
RfD	=	reference dose for non-carcinogenic effect of COC (mg/kg-day).

Applicable target risk limits (TR, THQ) for health protection can be matched to levels specified by the environmental regulatory authority. Toxicological parameters for each COC can be determined from published references, such as the U.S. EPA Integrated Risk Information System (IRIS). Exposure rates correspond to the chronic rate of contact or intake of the affected exposure medium (air, water, soil) by the receptor under anticipated land use conditions. As a conservative measure, these rates can be estimated based on standard exposure factors published by the regulatory authority or other source (e.g., American Industrial Health Council) for the anticipated land use at the site (e.g., residential, commercial, etc.).

Quantitative measures for derivation of RBELs for ecological receptors are not well defined. However, if the pathway screening evaluation indicates a reasonable potential for ecological exposure (e.g., surface water/aquatic species), applicable RBELs may be based on published standards or ecological screening criteria (e.g., surface water quality standard for aquatic life protection, ecological screening limits for terrestrial species, etc.). The U.S. EPA and various state agencies maintain databases of ecological screening levels for various types of receptors. However, given the highly conservative nature of these concentration limits, use of these values as ecological RBELs is appropriate only for preliminary screening-level analyses.

2.2.3 *Natural Attenuation Factor*

For each complete exposure pathway, the cumulative NAF represents the sum effect of various partitioning, dilution, and attenuation factors acting to reduce constituent concentrations during transport from source to receptor. These NAF components may involve both cross-media transfer factors (e.g., soil-to-air volatilisation, soil-to-groundwater leaching) and lateral transport factors (e.g., air dispersion, groundwater advection-dispersion). For exposure pathways with multiple POEs, separate NAF values must be derived for each POE location (e.g., ambient vapour inhalation by on-site worker and off-site resident; or groundwater ingestion at both hypothetical and actual wells). For a given site and exposure pathway, the NAF value may vary among evaluation of Tiers 1, 2, and 3, based on use of improved site data and evaluation methods.

For each complete exposure pathway and COC, the applicable NAF value can be derived based on either: i) the actual measured concentration ratio between the source medium and the POE or ii) fate-and-transport modelling analyses predicting this concentration ratio. For purpose of simplicity and accuracy, direct field measurements represent the preferred method of NAF estimation, whenever feasible. However, due to temporal variability and sampling difficulties, some of these factors can prove difficult to quantify via direct field measurements (e.g., soil volatilisation or leaching factors). In this case, modelling analyses, based on appropriate site-specific data and conservative assumptions, provide a convenient method of estimation. Groundwater dilution attenuation factors are amenable to direct measurement via wells spaced along the spine of the plume. In all cases, time-series groundwater monitoring data should be evaluated to establish the stability condition of the affected groundwater plume. Stable or diminishing plumes pose no risk to downgradient receptors located outside the plume area (i.e., NAF = infinite). Consequently, groundwater modelling analyses are necessary only for plumes for which available data are insufficient to establish the stability condition or indicate an expanding plume.

Various models are available to derive site-specific NAF values, ranging from simple analytical equations with limited data requirements (e.g., Tier 1 and Tier 2 evaluation) to complex numerical models requiring three-dimensional data resolution (e.g., Tier 3 evaluation). The appropriate evaluation tier and modelling tool will depend on the adequacy of the available data, the relative complexity of the site, the acceptable degree of conservatism, and the anticipated cost of the remedial action.

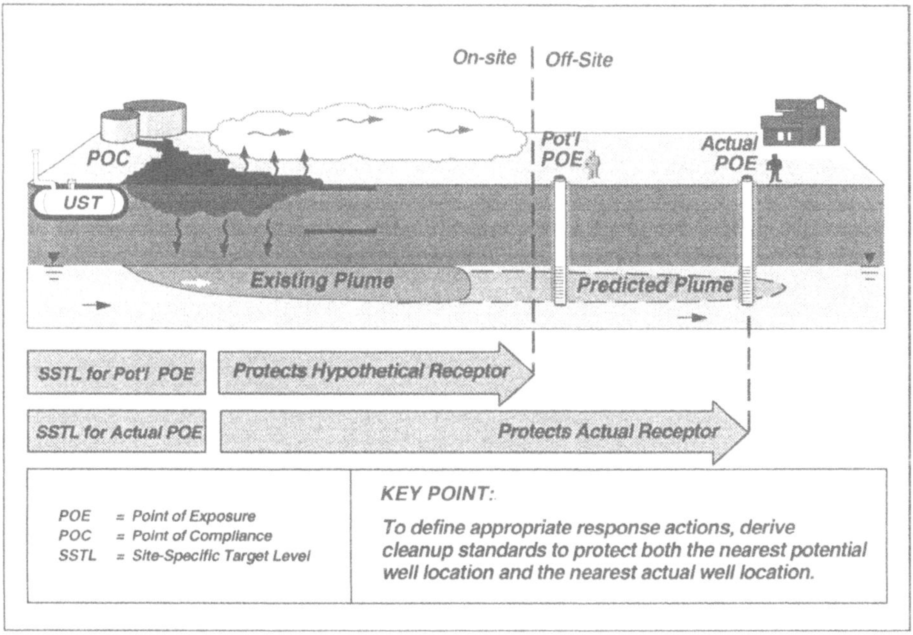

Figure 1. SSTL Analyses for Both Potential and Actual Receptors

2.2.4 Actual vs. Potential POEs

In development of SSTL values, careful distinction should be made between *actual* POEs, that correspond to existing conditions (e.g., current site users, existing water supply well), and *potential* POEs that represent possible future conditions (e.g., future users, hypothetical water supply well). Engineered remedies may be needed for protection of actual receptors, whereas, for potential receptors, sufficient time may be available for use of natural attenuation remedies. To accommodate this remedy selection strategy, each SSTL should be characterised as to the potential vs. actual nature of the exposure condition. For example, for surface soil exposure pathways, unpaved affected soil may pose a current exposure to on-site workers or residents, whereas soils beneath existing pavement represent a potential future exposure in the event the pavement were removed.

For the groundwater ingestion pathway, it is advisable to derive separate SSTL values both for a potential POE and an actual POE location as shown on Figure 1. For some programs, the location of the hypothetical groundwater ingestion POE may be specified under applicable regulations (e.g., nearest downgradient off-site property line). However, to support appropriate remedy selection, a second SSTL can be derived based on the nearest actual downgradient water supply well location.

2.2.5 *Risk Reduction Requirements*

The SSTL represents an action level for affected media in the source zone. Source media containing COC concentrations in excess of applicable SSTLs will require further assessment or remediation to control exposure via the relevant exposure pathway(s). If the SSTLs were estimated on the basis of limited site-specific data or highly conservative assumptions, the appropriate response may entail further site assessment and re-evaluation of appropriate target levels. For those pathways for which the results of the site-specific evaluation are reliable, appropriate exposure control measures must be selected and implemented, as discussed below.

2.2.6 *Applicable Data Evaluation Tools*

Derivation of SSTL values involves calculation of NAF values for each complete exposure pathway and relevant constituent of concern. Analytical models which can be used for estimation of steady-state NAF values for various air, soil, and groundwater exposure pathways under Tiers 1 and 2 are identified in Appendix X3 of RBCA Standard PS 104-98 [2]. As noted above, it is advisable to evaluate SSTLs for both *actual* and *potential* POE locations in order to support remedy selection. In addition, transient modelling analyses can provide important information regarding the timing and duration of the potential impact.

2.3 REMEDY SELECTION

For each exposure pathway determined to pose a health/environmental concern, a cost-effective remedy must be selected and implemented to achieve necessary risk reduction in the appropriate timeframe. This step of the site evaluation process involves development of an overall *exposure control strategy* and selection of optimal remediation technologies to implement this strategy.

The goal of the risk-based site management effort is to minimise risk by preventing exposure to harmful levels of site constituents. Risk reduction can be achieved by addressing any component of the exposure pathway: i) removing or treating the source, ii) interrupting contaminant transport mechanisms, or iii) controlling activities at the point of exposure. The remedial action plan may consist of one or more exposure control strategies, including:

- *Removal/Treatment Action*: Removal or treatment of affected source media (i.e., affected soils, groundwater, etc.) to reduce COC concentrations to levels less than or equal to applicable SSTLs (e.g., via excavation, soil venting, pump-and-treat, etc.).
- *Containment Measures*: Long-term engineering controls to prevent migration of harmful concentrations of COCs from the source to the POE (e.g., surface cover/capping, barrier walls, soil stabilisation, hydraulic containment, etc.).

- *Natural Attenuation Monitoring*: Periodic sampling and analysis to confirm stabilisation or reduction of affected media concentrations via natural attenuation processes.
- *Institutional Controls*: Legal or administrative measures to control the nature and frequency of human exposure at the POE (e.g., deed notice, alternative water supply, etc.).

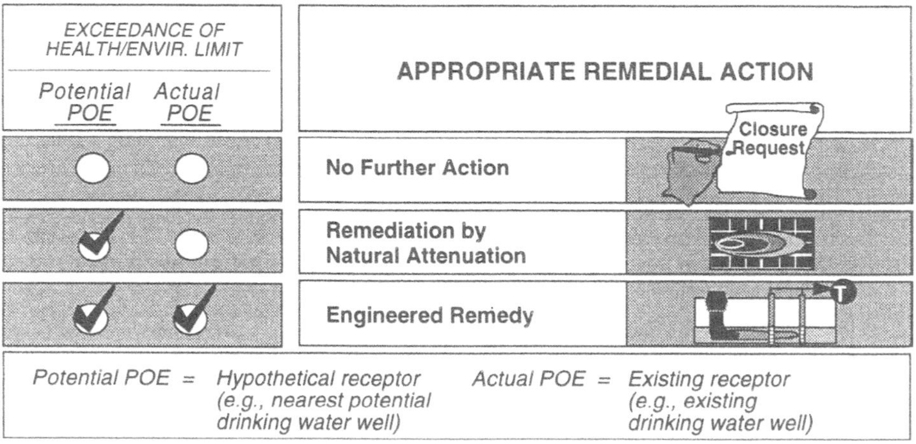

Figure 2. Risk-Based Remedy Selection Criteria

The appropriate exposure control strategy for a given site will depend on the nature of the risk reduction requirements. For example, as shown on Figure 2, engineered remedies (such as removal/ treatment or containment strategies) are appropriate for response to current or anticipated impacts on actual receptors. If risk reduction is required only for protection of potential future receptors (e.g., hypothetical water well users), groundwater remediation by natural attenuation may be employed to confirm plume stabilisation or reduction. No response action is required if constituent concentrations do not exceed SSTL values for either actual or potential receptors. The estimated *time to impact* determined in the risk-based site evaluation is also a key consideration in the remedy selection process. For example, if source media concentrations presently exceed an applicable SSTL value but the corresponding RBEL is not likely to be exceeded at the POE for an extended time period, additional time may be available for re-evaluation of potential exposure conditions based on site-specific monitoring program.

2.4 COMPLIANCE MONITORING PROGRAM DESIGN

Under many regulatory programs, a final compliance monitoring period will be required to confirm satisfactory completion of the remedy. Compliance monitoring (or *verification sampling*) typically involves sampling of one or more locations on an established schedule to identify either i) an exceedence of an applicable concentration limit or ii) a change of condition (e.g., change of land use, failure of engineering control) that might invalidate the basis for the remedy selection. If, upon completion of the monitoring period, compliance with applicable concentration limits is demonstrated, no further action is required.

Applicable Data Evaluation Tools: To confirm compliance with applicable cleanup standards, compliance monitoring *action levels* for the groundwater exposure pathway can be derived using the same models used for SSTL calculation. Under this approach, groundwater compliance monitoring locations are selected between the source location (point of compliance) and the point of exposure (POE). For each monitoring point, a NAF value is calculated for constituent transport from the monitoring point to the POE, and the action level is calculated as the arithmetic product of the NAF times the applicable RBEL for each constituent. If action levels are exceeded during the compliance monitoring period, further evaluation may be required to ensure adequate protection of downgradient receptors.

3. Model Selection Guidelines

Under the risk-based site evaluation process outlined above, fate-and-transport models are used to derive SSTL values based on estimation of the pathway-specific natural attenuation factor (NAF). Whenever feasible, direct field measurements (e.g., soil-to-air based on time-series monitoring) represent the preferred method of NAF estimation. However, if the exposure pathway is not amenable to direct NAF measurement (e.g., volatilisation factors, leachate factors, etc.) or if time-series analyses show the contaminant zone to be expanding over time, modelling analyses, based on appropriate site-specific data and conservative assumptions, can provide a convenient method of estimating future exposure levels.

The "best" model for a given site will be the simplest model providing a reliable and reasonably conservative prediction of potential exposure. Under the ASTM RBCA process, relatively simple analytical modelling tools are applied under Tiers 1 and 2, followed by more sophisticated modelling methods, if warranted, under Tier 3. The choice between simple and complex modelling methods should be dictated by the adequacy of the site database and the relative degree of error likely to be introduced by the model itself. In addition, the cost of upgrading to a more complex Tier 3 site evaluation must be warranted by the potential reduction in site remediation costs or the complex nature of the anticipated exposure condition. General guidelines for application of various types of fate-and-transport models under Tiers 1, 2 and 3 are summarised below.

3.1 MODEL DIMENSIONS

For each of the exposure pathways addressed in the RBCA standard, fate-and-transport models are available to estimate NAF values based on either a one-, two-, or three-dimensional analysis of contaminant transport. In reality, all contaminant transport occurs in three dimensions; however, one-dimensional (1-D) or two-dimensional (2-D) modelling tools may be employed for purpose of conservatism and simplicity. One-dimensional models, which ignore lateral and vertical dispersion effects, may significantly overestimate exposure levels and underestimate the pathway NAF. For this purpose, 2-D fate-and-transport models are commonly employed for Tier 1 and Tier 2 analyses, as presented in Appendix X3 of the ASTM RBCA Standard. Three-dimensional transport models may provide a more accurate and less conservative NAF estimate under a Tier 3 evaluation, but must be supported by three-dimensional measurements of key transport parameters (e.g., hydraulic conductivity, etc.).

3.2 STEADY-STATE VERSUS TRANSIENT ANALYSES

Steady-state fate-and-transport models, which assume a constant source concentration and constant flow conditions over time, provide a conservative (lowerbound) NAF estimate corresponding to maximum chronic exposure conditions. In reality, following termination of the release, source concentrations in soil and groundwater are likely to diminish over time, resulting in time-variable exposure concentrations at the POE. For purpose of simplicity and conservatism, steady-state, constant-source models, providing a lowerbound NAF value, are commonly employed under Tiers 1 and 2. However, to support risk management decisions, these constant-source models can be run in a transient mode to predict the *time to impact*, i.e., the time required for the exposure concentration to exceed the RBEL at the POE. Under Tier 3, fully transient models, simulating both time-variable source concentrations and transport phenomena, can be used to characterise both the timing and *duration* of the RBEL exceedence. Again, these more sophisticated Tier 3 analyses must be based on sufficient site-specific data to support reliable modelling results.

3.3 PROBABILISTIC VERSUS DETERMINISTIC MODELS

Under Tiers 1 and 2, exposure concentrations and NAF values are characterised on the basis of *deterministic* models which provide a unique output value for each unique set of input values. Uncertainty in the modelling analysis is addressed by means of a *sensitivity study*, i.e., by varying key input values to evaluate their potential impact on the model output. Under Tier 3, *probabilistic* modelling may be employed as a more sophisticated approach to management of model uncertainty. In probabilistic modelling, for each key input parameter, the user provides a probability distribution corresponding to the range and type of distribution observed for the parameter at the site. The model then completes the fate-and-transport calculation for the full range of these input values, effectively conducting multiple random model sensitivity studies. The model result is not a unique value but a probability distribution defining the possible range of results (e.g., exposure concentration, NAF value) for the specified site conditions. The probabilistic analysis provides the user with relatively sophisticated information regarding possible exposure conditions (e.g., for a given SSTL value, what is the probability that the RBEL will be exceeded at any future time?) However, to support reliable results, this Tier 3 modelling method will typically require significant additional site characterisation data relative to Tier 1 or Tier 2 deterministic analyses.

4. RBCA Tool Kit for Chemical Releases

The *RBCA Tool Kit for Chemical Releases* has been developed by Groundwater Services, Inc. (GSI), expressly for use with the Tier 1 and Tier 2 site evaluation procedures outlined in the ASTM RBCA Standard [4]. Based upon site-specific data supplied by the user, the *RBCA Tool Kit* combines fate-and-transport modelling and risk characterisation functions to compute the following:

- Exposure Concentrations
- Average Daily Intake
- Baseline Risk Levels
- Risk-Based Media Cleanup Levels

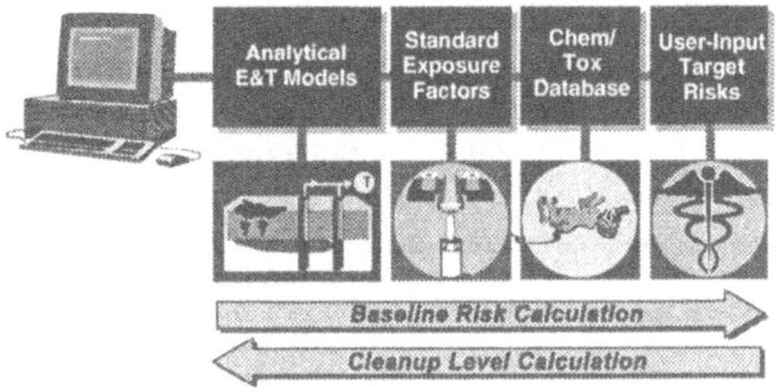

Figure 3. RBCA Tool Kit

Key *RBCA Tool Kit* features relevant to SSTL calculations and risk-based remedy selection are outlined below.

4.1 MODEL CALCULATION FUNCTIONS

Using a system of ten analytical models linked to internal libraries of standard exposure factors and chemical/toxicological data for over 90 compounds, the *RBCA Tool Kit* can calculate either baseline risk levels or cleanup standards for each complete exposure pathway identified by the user. Key calculation steps are as follows:

4.1.1 Exposure Concentrations

Based on representative concentrations of constituents of concern (COCs) present in the affected source media, maximum steady-state concentrations likely to occur at the point of exposure (POE) are calculated using the steady-state analytical fate-and-transport models identified in Appendix X3 of ASTM PS 104-98 [2]. To perform these calculations, the system evaluates cross-media partitioning (e.g., volatilisation from soil to air) and lateral transport from the source to the POE (e.g., contaminant transport via air or groundwater flow). The source media and optional exposure pathways included in the software are as shown in table 1.

TABLE 1. Source media and exposure pathways

SOURCE MEDIA	EXPOSURE PATHWAYS
Surface Soils	Inhalation of Vapour and Particulates
	Dermal Contact with Soil
	Ingestion of Soil and Dust
	Leaching to Groundwater
Subsurface Soils	Inhalation of Vapour
	Leaching to Groundwater
Groundwater	Ingestion of Potable Water
	Inhalation of Vapour
	Discharge to Surface Water
	• Ingestion/Dermal Contact via Swimming
	• Ingestion via Fish Consumption
	• Aquatic Life Protection

4.1.2 Average Daily Intake

Based upon the exposure factors selected by the user, the average daily chemical intake for each receptor along each selected pathway is calculated in accordance with EPA guidelines [4]. These values are used in baseline risk calculations for each complete pathway.

Baseline Risk Characterisation: Human health risks associated with exposure to COCs are calculated by the software on the basis of average daily intake rates and the corresponding toxicological parameters for carcinogenic and non-carcinogenic effects. For each complete pathway, the system output provides both individual and additive constituent results for carcinogens and non-carcinogens.

Media Cleanup Values: The *RBCA Tool Kit* has the ability to i) compare the site data to Tier 1 Risk-Based Screening Levels (RBSLs), computed using the default parameter values as listed in ASTM PS 104-98, or ii) calculate Tier 2 Site-Specific Target Levels (SSTLs) based on user-supplied site information. For each source medium (i.e., affected soil and groundwater), the software reports target concentrations for all complete pathways and identifies the applicable (i.e., minimum) value for source remediation.

4.2 RISK-BASED DECISION SUPPORT FEATURES

The *RBCA Tool Kit for Chemical Releases* includes several features designed to support key steps of the risk-based site evaluation process, including the following:

4.2.1 Step-by-Step Evaluation Process

From the Main Screen of the graphical user interface, the user is guided through all the necessary steps for completing the Tier 1 or Tier 2 evaluation process. On subsequent screens the interface leads the user through exposure pathway identification, model selection, site-specific parameter input, and output review. All output screens may be printed in a report-quality format.

4.2.2 Analysis of Actual and Potential POEs

Multiple off-site exposure points are allowed for the groundwater and outdoor air pathways. This enables the user to evaluate risks at both actual (e.g., an actual nearby well) and potential (e.g., a hypothetical well at the property boundary) POEs. Whether site risks affect an actual or potential POE adds a qualitative dimension to the risk calculations, which may be an important factor in remedy selection at some sites.

4.2.3 Transient Groundwater Modelling Analyses

An optional Transient Domenico Worksheet is provided to allow the user to estimate the time required for site constituents to impact off-site groundwater POEs. Groundwater risk levels and cleanup standards calculated by the software are based on steady-state concentrations. However, the time to reach steady-state concentrations at off-site POEs may be very long for some constituents. Thus, the time required to exceed a concentration limit at a POE may be an important factor in remedy selection as near-term impacts may require a significantly different response than longer-term impacts (e.g., an engineered response versus natural attenuation).

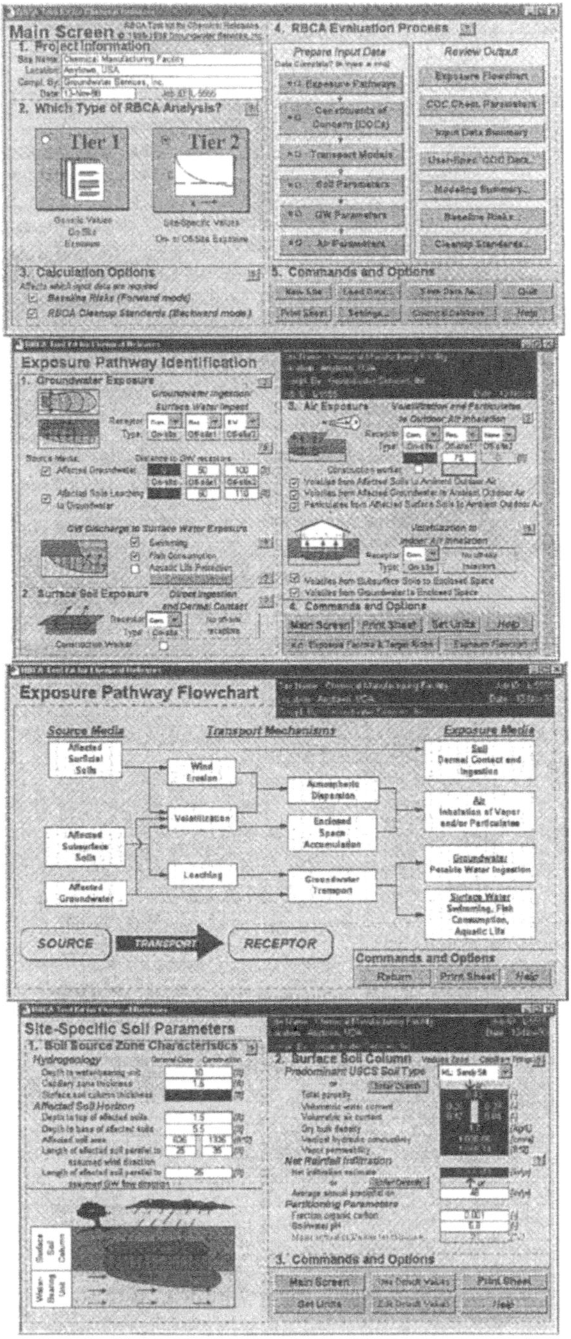

Figure 4. Example screens from the RBCA Tool Kit for Chemical Releases

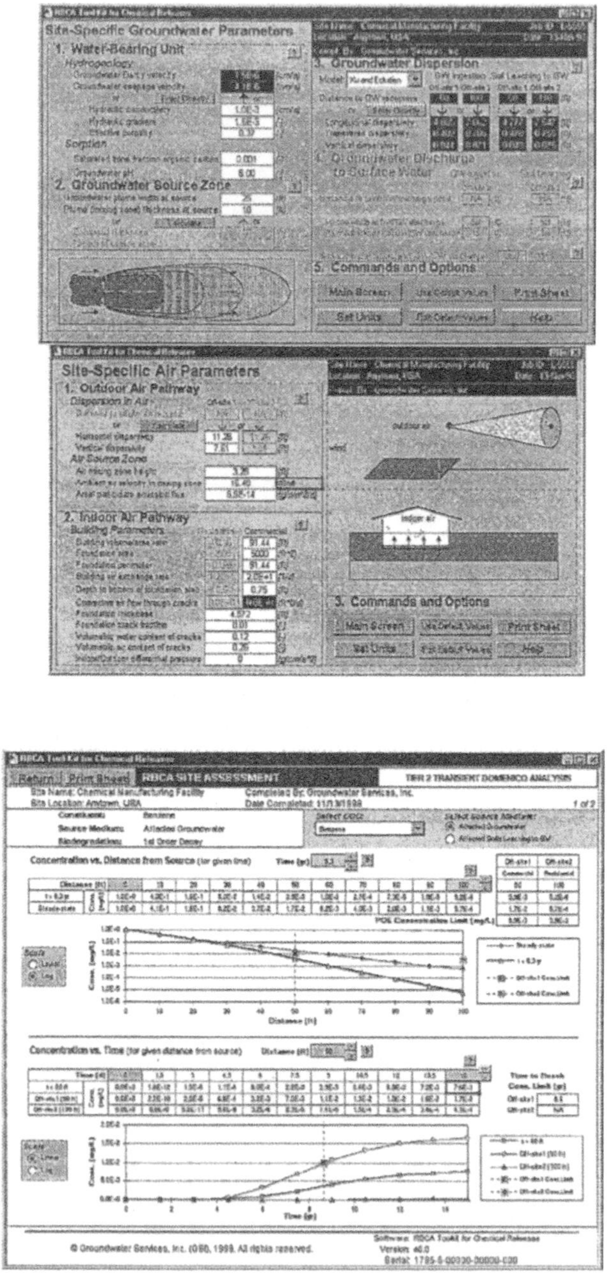

Figure 4. Continued Example screens from the RBCA Tool Kit for Chemical Releases

5. Conclusion

As demonstrated by the *RBCA Tool Kit for Chemical Releases*, a system of simple analytical fate-and-transport models can be used for comprehensive risk-based evaluation of potential soil, air, groundwater, and surface water exposure pathways. However, as with all predictive modelling efforts, reliable results require proper characterisation of site-specific input parameters. In all cases, model predictions must be shown to be consistent with the actual constituent distributions observed at the site. Use of the Tier 1 and Tier 2 calculation methods outlined in the ASTM RBCA Standard and incorporated in the *RBCA Tool Kit* can significantly reduce the time and effort required for evaluation of risk reduction requirements and selection of appropriate exposure control methods. However, proper scientific and/or engineering expertise is required both for characterisation of input parameters and assessment of model results.

6. References

1. ASTM, (1995), *Standard Guide for Risk-Based Corrective Action at Petroleum Release Sites,* ASTM E-1739-95, Philadelphia, PA.

2. ASTM, (1998), *Standard Provisional Guide for Risk-Based Corrective Action,* ASTM PS 104-98, Philadelphia, PA.

3. ASTM, (1998), *Standard Guide for Remediation of Ground Water by Natural Attenuation at Petroleum Release Sites,* ASTM E-1943-98, Philadelphia, PA.

4. Connor, J.A., R.L. Bowers, J.P. Nevin, (1998), *User's Guide: RBCA Tool Kit for Chemical Releases,* Groundwater Services, Inc., Houston, Texas.

5. J.P. Nevin., McHugh, T.E., C.J. Newell, J.A. Connor, N.J. Novick, (1998), *Practical Methods for Demonstration of Groundwater Remediation by Natural Attenuation (RNA),* Proceedings of NGWA Petroleum Hydrocarbon Conference, Houston, Texas.

DANISH EPA USE OF MODELS FOR ASSESSMENT OF PESTICIDES MOBILITY

CHRISTIAN DEIBJERG HANSEN
Ministry of Environmental and Energy
Danish EPA, Pesticide Division
Strandgate 29,
1451 Copenhagen K,
Denmark

Abstract

The Danish EPA have increasingly used models in the assessment of pesticides mobility. They have set up certain requirements for a model study under Danish conditions. These requirements are printed on a CD-rom, which Danish EPA released last year. The CD includes the input files for the model MACRO, such as climate data and the parameter file for two scenarios. It furthermore includes a thoroughly description of the scenarios (soil types, water balances, climate data etc.). The scenarios represent a typical Danish sandy soil and a loamy clay soil.
A study concludes that models which not include a description of preferential flow are not feasible for prediction of leaching in macroporous soils. Therefore, the Danish EPA have decided to use the Swedish model MACRO. Since the model is not validated, a realistic worst case must be covered with respect to pesticide parameters, application rate, crop, etc. A study has shown that hourly precipitation data can give up to 12 times higher leaching, than using daily precipitation data.

1. Introduction

The Danish EPA have increasingly used mathematical models in the assessment of whether a pesticide presents a risk of groundwater contamination or not. The reason is that ordinary documentation in the form of laboratory tests in many cases is not sufficient to evaluate the risk of groundwater contamination. Nor does supplemental documentation in the form of lysimeter studies always provide a definitive description of the actual risk in the field.

J.B.H.J. Linders (ed.), Modelling of Environmental Chemical Exposure and Risk, 183–192.

2. Transport of Pesticides

Using models to describe transport of pesticides in soils allows testing of many more combinations of factors, such as soil type, climatic conditions, application times, anticipated soil exposure, etc. than is possible otherwise financially and technically in lysimeter and field testing.

Modelling is increasingly being used as documentation, and are typically included in EU monographs. It can be very difficult to assess model studies since models often include a wide range of adjustable parameters. For some of them a small adjustment can hugely affect the simulated leaching. Therefore, the Danish EPA have set up certain requirements for a model study, so it is possible to make equal assessments, and in that way decide whether or not, a pesticide will leach under Danish conditions. These requirements are printed on a CD-rom, which the Danish EPA released last year [4]. The CD-rom include the input files for the model MACRO, such as climate data and the parameter file for two scenarios. Furthermore, the CD-rom includes a detailed description of the scenarios (soil types, water balances, climate data, etc.) and accompanying documentation of the calibrated water balances.

3. Scenarios

The two scenarios the Danish EPA have selected are two actual localities in Denmark; Karup and Langvad. These two localities were chosen because they are well-described with respect to climate and soil types. Data also exists in the form of field measurement of water balances. Furthermore the two scenarios Karup (light soil with coarse sand) and Langvad (loamy clay soil with macropores) are also soils considered to represent the soil types, found most often in Denmark, sand and loam. Both localities are in areas receiving more precipitation than in average for either sandy soil and loam, which therefore, with respect to climate make the scenarios realistic worst cases.

3.1 SOIL

Soil hydraulic parameters in terms of retention curves and saturated hydraulic conductivity were originally measured using soil samples from two experimental sites within the catchments - from Rabis creek in the Karup catchment and from Syv creek in the Langvad catchment [7]. The water balance and nitrate transport was then simulated on plot scale using the Daisy model prior to the catchment modelling with Mike She. The physical parameters representing the two locations are shown in Tables 1 and 2. Each soil profile was devided into three horizons with respect to hydraulic parameters and in four horizons with respect to organic matter content. Retention curves (measured and as represented in MACRO) are shown in Figures 1 and 2.

TABLE 1. Soil physical and chemical characteristics for the sandy soil in Karup

Karup		Soil texture (%)			Physical characteristics			
Depth (cm)	Clay	Silt	Fine sand	Coarse sand	Bulk density (g/cm^3)	θ_s(%)	K_s (mm/h)	Org. C (%)
0-20	6.0	7.0	21.0	66.0	1.51	42	216	2.20
20-40	5.5	5.5	14.0	75.0	1.55	42	125	1.74
40-70	5.5	5.5	14.0	75.0	1.55	42	125	0.30
70-	4.2	4.8	16.0	75.0	1.53	42	721	0.20

K_s: Saturated hydraulic conductivity, θ_s: Total effective porosity.

TABLE 2 Soil physical and chemical characteristics for the loamy clay soil in Langvad

Langvad		Soil texture (%)			Physical characteristics			
Depth (cm)	Clay	Silt	Fine sand	Coarse sand	Bulk density (g/cm_3)	θ_s (%)	K_s (mm/h)	Org C (%)
0-20	16.0	22.0	40.2	21.8	1.50	45	90	2.10
20-40	20.0	21.0	40.0	19.0	1.65	36	360	1.42
40-70	20.0	21.0	40.0	19.0	1.65	36	360	0.50
70-	20.6	20.0	41.0	18.4	1.76	34	26	0.26

K_s: Saturated hydraulic conductivity, θ_s: Total effective porosity.

For both soil types, the conceptual model describes a 2.5 m deep soil column containing three different horizons. For the sandy soil the lower boundary is defined as a constant potential close to field capacity. This approach mimics the free drainage conditions in a deep unsaturated profile overlying a sandy aquifer. This soil type was considered to be rather homogeneous, and macropore flow was not considered. On the loamy soil, the lower boundary was defined as having a constant gradient which was calibrated in order to ensure realistic fluctuation of the groundwater table. This approach mimics the situation with a shallow topsoil overlaying an aquifer in 2.5 m depth. Additionally, artificial drains were applied in 1.3 m depth, and macropore flow was considered, with the macroporosity decreasing from 2 % in the top soil to 1 % in the deepest horizon.

3.2 WATER BALANCE

The water balances shown in the following were calibrated using MACRO version 4.0.

The annual water balance representing the Karup location was calibrated for the period 1983-1985 using simulated values of evapotranspiration and percolation for a spring cereal obtained with the Mike She model established and validated in a former project.

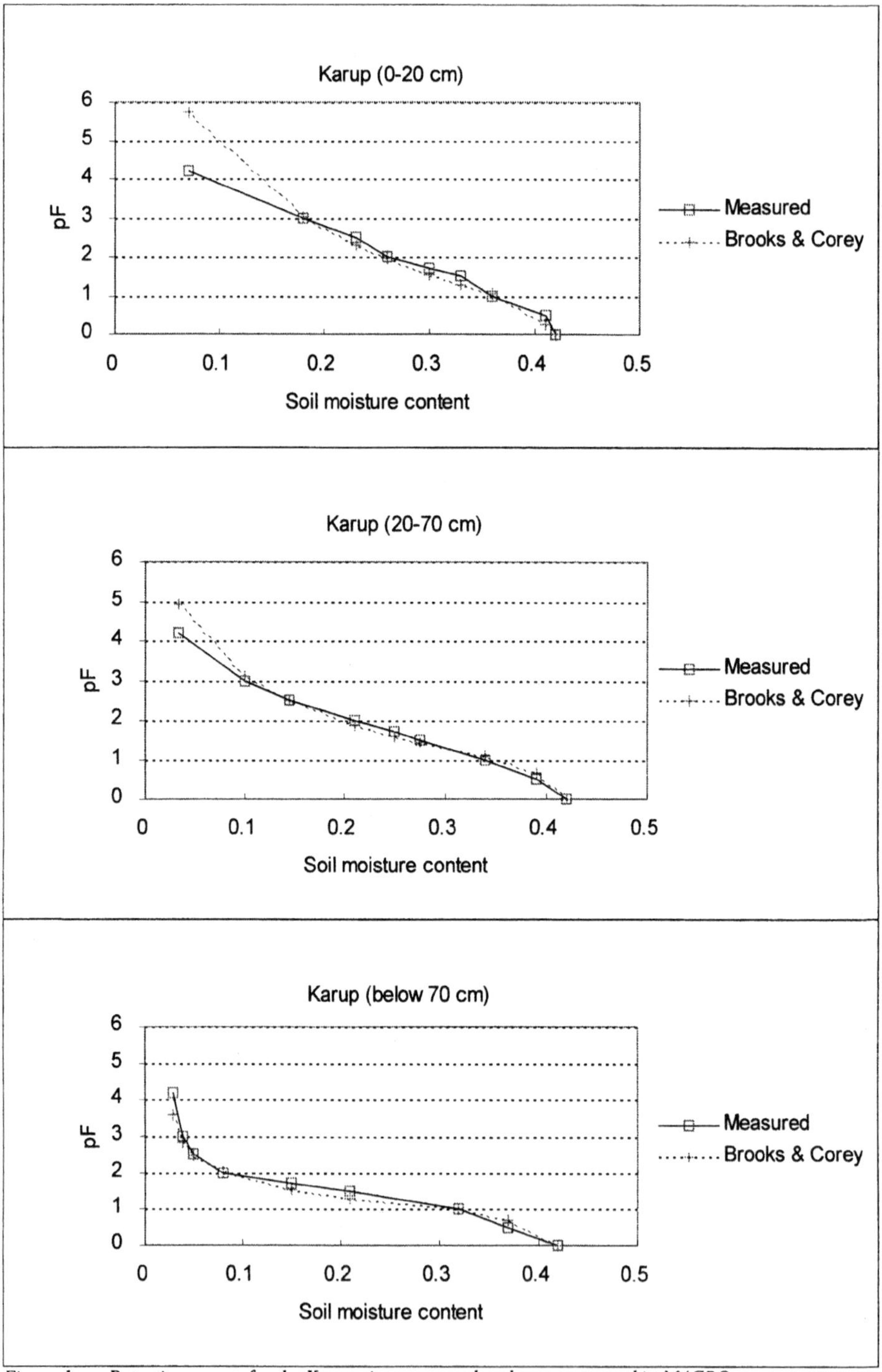

Figure 1 Retention curves for the Karup site, measured and as represented in MACRO.

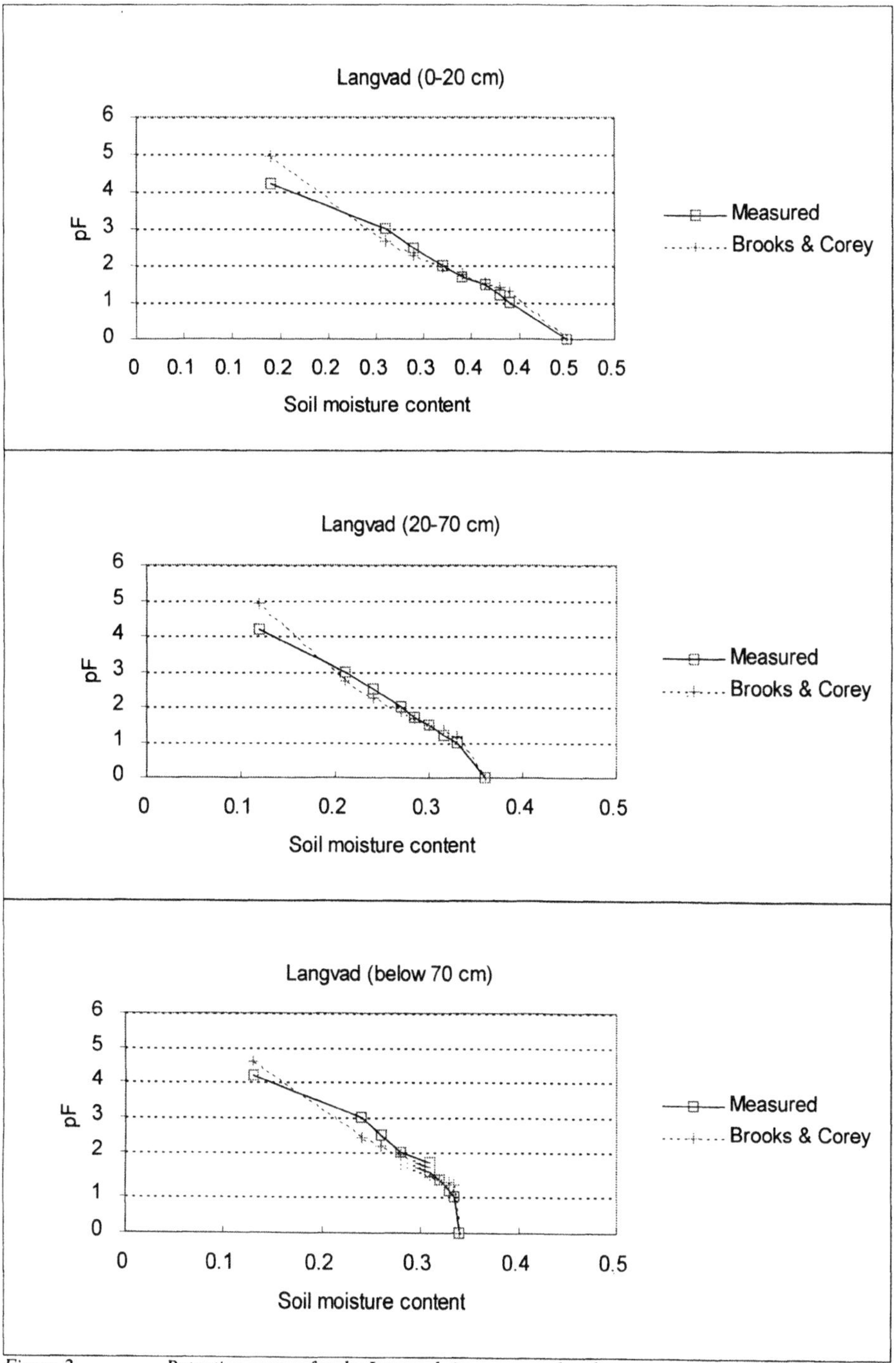

Figure 2 *Retention curves for the Langvad site, measured and as represented in MACRO.*

TABLE 3 *Calibrated annual water balance for the Karup site*

Year	Precipitation (mm)	Actual evapotranspiration (mm)	Percolation in 2.5 m depth (mm)
1983	1017	451	586
1984	855	418	455
1985	927	485	388

The annual water balance representing the Langvad location was calibrated for the period 1983-1985 using the experienced balances in terms of evapotranspiration, drain flow and percolation for a spring cereal obtained in a former project.

TABLE 4 *Calibrated annual water balance for the Langvad site*

Year	Precipitation (mm)	Actual evap. (mm)	Drainage (mm)	Percolation in 2.5 m depth (mm)
1983	626	442	81	140
1984	680	443	24	170
1985	689	461	52	146

3.3 CLIMATE

Daily climatic data in terms of precipitation, minimum and maximum air temperatures and sunshine hours representing the two locations were obtained from Danish Meteorological Institute. The precipitation data were afterwards corrected for wind and wetting losses according to Allerup and Madsen [1], and in addition it was ensured that the annual rainfall complied with the areal mean values which were originally used for the water balance calculations in Storm *et al.* [7]. Potential evapotranspiration was calculated using a modified Makkink approach [2] with the use of radiation data calculated from sunshine hours using the Ångström formula [6]. Minimum and maximum air temperatures were used directly. In Table 5 annual precipitation and potential evapotranspiration for the two locations are shown.

3.4 INPUT PARAMETER

Since models not are validated and probably never will be, it is important that the substance specific parameters, application rate and crop parameters should be chosen in such way that they represent a realistic worst case, to meet the criterion that the assessment should be conservative.

A combination of degradation rate and sorption conditions can be chosen, e.g. on the basis of sensitivity analysis. The combination must however, be chosen as starting point among the lowest Koc values and highest DT50 values.
Application must be the highest recommended dose, on the worst case crop (lowest plant cover and plant uptake, etc.). Several simulations must be done with changing application dates within the application window, to catch eventual extreme events.
In a rapport [3], the Danish Geological Survey (GEUS) have tried to quantify the variation in leaching, then changing input parameters, using the MACRO 4.0 model. The study tested the possible difference in the leaching, using an hourly versus a daily time step for the precipitation as input.

TABLE 5. Climatic data available for Karup and Langvad

Year	Karup		Langvad	
	Precipitation (mm/year)	Pot. evap. (mm/year)	Precipitation (mm/year)	Pot. evap. (mm/year)
1968	895	611	-	-
1969	647	593	-	-
1970	848	609	-	-
1971	702	625	-	-
1972	714	609	-	-
1973	650	625	-	-
1974	862	656	795	645
1975	694	649	622	677
1976	640	632	609	666
1977	941	613	606	635
1978	811	574	857	612
1979	939	566	662	592
1980	1064	577	805	604
1981	1078	583	769	618
1982	891	614	598	667
1983	1017	612	626	630
1984	855	607	680	619
1985	927	600	689	604
1986	904	598	620	611
1987	1044	595	692	575
1988	992	607	639	623
1989	828	645	562	648
1990	1053	627	650	661
1991	901	642	697	628
1992	941	685	509	669
1993	855	615	818	626
1994	1062	627	744	662
1995	799	634	533	660
1996	590	614	424	644

- Precipitation data have been corrected for wind and wetting losses [1]
- Potential evapotranspiration is based on a modified Makkink equation [2]

The study showed that using hourly time step, can lead to 12 times higher annually leaching . The Danish EPA accept a daily time step, but will as a consequent of the study consider changing to hourly in the future. There are, however, arguments against using an hourly time step for precipitation. One is that it is not possible to get time series, which are more than 7 to 8 years long. Daily time step series are available for periods up to 30 years, which therefore cover a more wide range of dry and wet years.

TABLE 6. Proportional relations between average concentrations of an hourly and dailybased precipitation. Spring application with two different pesticides (K and M) and tree different precipitation data (Rain1..3) [3].

	Spring											
	Karup						Langvad					
	K			M			K			M		
Year	Rain1	Rain2	Rain3	Rain1	Rain2	Rain3	Rain1	Rain2	Rain3	Rain1	Rain2	Rain3
1988	-	-	-	-	-	-	-	-	-	-	-	-
1989	1.0	6.4	1.0	1.0	6.4	1.0	1.0	7.1	1.1	1.0	6.2	1.0
1990	1.0	1.0	1.0	1.0	1.0	1.0	1.0	1.1	1.1	1.0	1.0	1.0
1991	4.3	9.0	1.0	4.3	9.2	1.0	3.4	12.3	1.1	3.1	8.1	1.0
1992	1.0	1.0	1.0	1.0	1.0	1.0	1.0	1.0	1.0	1.0	1.0	1.0
1993	1.0	1.0	1.0	1.0	1.0	1.0	1.1	1.1	1.1	1.0	1.0	1.0
1994	6.4	1.7	7.4	6.5	1.7	7.4	7.2	2.5	8.1	7.1	2.5	8.3
1995	1.0	1.0	1.6	1.0	1.0	1.6	1.0	1.2	2.2	1.0	1.0	2.2

TABLE 7. Proportional relations between average concentrations of an hourly and daily based precipitation. Autumn application with two different pesticides (K and M) and tree different precipitation data (Rain1..3). [3]

	Autumn											
	Karup						Langvad					
	K			M			K			M		
Year	Rain1	Rain1	Rain1	Rain1	Rain1	Rain1	Rain1	Rain1	Rain1	Rain1	Rain1	Rain1
1988	0.9	3.3	1.0	1.0	0.9	1.0	0.6	2.2	1.3	0.9	3.1	0.9
1989	3.5	1.0	2.9	2.9	3.5	2.9	3.5	0.8	2.6	3.6	1.0	1.3
1990	1.0	1.0	1.8	1.8	1.0	1.8	0.9	0.8	1.6	1.0	1.0	0.9
1991	1.6	1.6	3.1	3.1	1.6	3.0	1.9	2.0	2.2	2.2	2.2	1.1
1992	4.2	4.9	4.6	4.6	4.2	4.5	3.8	4.4	4.7	4.4	4.8	1.8
1993	1.2	2.3	2.2	2.2	1.2	2.1	1.2	2.7	2.4	1.2	2.3	2.0
1994	4.5	1.0	1.7	1.7	4.5	1.7	4.1	0.7	2.4	4.5	1.1	-
1995	1.0	1.0	1.0	1.0	1.0	1.0	0.8	1.2	0.9	1.0	1.0	-

3.5 CHOICE OF MODEL

A criterion for THE choice of model code should be that preferential flow is incorporated in the model. A study financed by the Danish EPA [8], tested 4 different model codes, PESTLA, PELMO, MACRO and Mike She. In a validation procedure

(FOCUS 1995) the model codes including a description of preferential flow processes (MACRO and Mike She) required less calibration efforts in order to meet the selected FOCUS performance criteria than those without such description (PELMO and PESTLA).

TABLE 8. Evaluation of model performance from experiments with soil columns and lysimeter simulations. The validation level is indicated by the F-test. The levels accepted in this test were F=2 or F=5.

	Experiment with soil column			
	Percolation, litre	PFBA tracer, mg	MCPP leaching, mg	F-test
Observed value	34.7	24	13.43	
PELMO	34	9.6	0.022	Neg
PESTLA	27.93	48.86	0.045	Neg
MACRO	28.02	6.47	8.58	F=2

	Experiment with lysimeters			
	Percolation, litre	Bromide tracer, mg	MCPP leaching, mg	F-test
Observed values				
Lysimeter 1	222	140	2.05	
Lysimeter 2	257	185	4.38	
PELMO	215	241	0.025	Neg
PESTLA	208	276	0.0001	Neg
MACRO	221	91	7.53	F=2

The study concludes that the two models (PESTLA and PELMO) codes not are feasible for prediction of contaminant leaching in macroporous soils. The Danish EPA have therefore decided to use the MACRO model.

Since mathematical models are still not validated properly, the use of models should be taken with cautions. Therefore it shall be stressed that models should not at present, be used as a substitute for experimental data, but rather as complementary.

4. References

1. Allerup, P. and H. Madsen, (1979): Accuracy of point precipitation measurements. Danish Meteorological Institute. Climatological Papers No. 5. Copenhagen.

2. Aslyng, H.C. and S. Hansen, (1982). Water balance and crop production simulation. Hydrotechnical Laboratory, The Royal Veterinary and Agricultural University, Copenhagen.

3. Barlebo, H.C. *et. al.* (1999): Udvaskning af pesticider: Modellering af betydning af variationen i danske jordarter, nedbør og stofparametre. (in Danish)

4. Danish EPA (1998): Input files for the macro model. Released as a CD-rom

5. FOCUS (1995): Leaching models and EU registration. Final report of the workgroup-FOrum for the Co-ordination of pesticide fate models and their Use, fundes by the European Crop Protection Association and COST Action 66. DOC. 4952/VI/95

6. Iqbal, M., (1983). An introduction to solar radiation. Academic Press.

7. Storm, B., M. Styczen and T. Clausen, (1990). Regional Model for Nitrate Transport and Transformation. *NPo Research in the NAEP*, Report B15. National Agency of Environmental Protection, Copenhagen (in Danish).

8. Thorsen, M. *et al.* (1998): Validation and Development of Pesticide leaching Models. Danish EPA pesticides Research no. 47.

AN OPTIMIZATION MODEL FOR THE CONTROL OF REGIONAL AIR QUALITY IN EUROPE

MARKUS AMANN, CHRIS HEYES, MAREK MAKOWSKI,
WOLFGANG SCHÖPP
International Institute for Applied Systems Analysis (IIASA)
A-2361 Laxenburg, Austria
E-mail: amann@iiasa.ac.at

Abstract

Cost-optimization appears as an attractive feature for the management of regional air quality. The paper presents the formulation of an optimization problem for an integrated assessment model dealing with regional air pollution in Europe. Putting constraints (targets) on regional air quality (for ground-level ozone exposure, acidification and eutrophication), the optimization identifies the cost-minimal combination of emission reduction measures at the various emission sources. The paper presents a practical example of an optimized scenario, as it is currently used in international negotiations in Europe and explores the cost-savings achieved by the optimization approach.

1. Introduction

There is substantial concern about the environmental impacts of air pollution on the local, regional and global scale. It has been shown that observed levels of various air pollutants can threaten human health, vegetation, wild life, and cause damage to materials. In order to limit the negative effects of air pollution, measures to reduce emissions from a variety of sources have been initiated.

Once emitted, many air pollutants remain in the atmosphere for some time before they are finally deposited on the ground. During this time, they are transported with the air mass over long distances, often crossing national boundaries. As a consequence, at a given site the concentration of pollutants and their deposition on the ground is influenced by a large number of emission sources, frequently in many different countries. Thus, action to efficiently abate air pollution problems has to be coordinated internationally.

J.B.H.J. Linders (ed.), Modelling of Environmental Chemical Exposure and Risk, 193–203.

Most of the current international agreements determine required abatement measures solely in relation to technical and economic characteristics of the sources of emissions, such as available abatement technologies, costs, historic emission levels, etc. No relation is established to the actual environmental impacts of emissions. For achieving overall cost-effectiveness of strategies, however, the justification of potential measures in relation to their environmental benefits must also be taken into account. Recently, progress has been made in quantifying the environmental sensitivities of various ecosystems. Critical loads and critical levels have been established reflecting the maximum exposure of ecosystems to one or several pollutants not leading to environmental damage in the long run. In Europe such threshold values have been determined, focusing on acidification and eutrophication as well as on vegetation damage from tropospheric ozone. The World Health Organization has established air quality guidelines for the protection of human health.

It is generally expected that the current policies on emission reductions will greatly reduce the levels of tropospheric ozone. However, the measures will not be sufficient to eliminate the problem everywhere in Europe. To meet full protection of human health and vegetation, additional measures will be necessary. Since most of the low-cost options for abating emissions are already adopted in the current strategies, further action aiming at the sustainability of Europe's ecosystems will have to embark on more costly measures. Cost-effectiveness will be an important argument for gaining acceptance of proposed policies.

2. An Integrated Assessment Models for Regional Air Pollution

The Regional Air Pollution INformation and Simulation (RAINS)-model developed at the International Institute for Applied Systems Analysis (IIASA, Laxenburg, Austria) provides a consistent framework for the analysis of emission reduction strategies, focusing on acidification, eutrophication and tropospheric ozone. RAINS comprises modules for emission generation (with databases on current and future economic activities, energy consumption levels, fuel characteristics, etc.), for emission control options and costs, for atmospheric dispersion of pollutants and for environmental sensitivities (i.e., databases on critical loads). In order to create a consistent and comprehensive picture of the options for simultaneously addressing the three environmental problems (acidification, eutrophication and tropospheric ozone), the model considers emissions of sulfur dioxide (SO_2), nitrogen oxides (NO_x), ammonia (NH_3) and volatile organic compounds (VOC). A detailed description of the RAINS model can be found in Alcamo *et al.* [1]. A schematic diagram of the RAINS model is displayed in Figure 1.

The European implementation of the RAINS model incorporates databases on energy consumption for 38 regions in Europe, distinguishing 22 categories of fuel use in six economic sectors. The time horizon extends from the year 1990 up to the year 2010. Emissions of SO_2, NO_x, NH_3 and VOC for 1990 are estimated based on information collected by the CORINAIR'90 inventory of the European Environmental Agency [4] and on national information. Options and costs for controlling emissions of the various substances are represented in the model by considering the characteristic technical and economic features of the most important emission reduction measures. Atmospheric dispersion processes over Europe for sulfur and nitrogen compounds are modeled based on results of the European EMEP model developed at the Norwegian Meteorological Institute [2]. For tropospheric ozone, source-receptor relationships between the precursor emissions and the regional ozone concentrations are derived from the EMEP photo-oxidants model [5], [8]. The RAINS model incorporates databases on critical loads and critical levels compiled at the Coordination Center for Effects (CCE) at the National Institute for Public Health and Environmental Protection (RIVM) in the Netherlands [7].

The RAINS model can be operated in the 'scenario analysis' mode, i.e., following the pathways of the emissions from their sources to their environmental impacts. In this case the model provides estimates of regional costs and environmental benefits of alternative emission control strategies. Alternatively, an 'optimization mode' is available to identify cost-optimal allocations of emission reductions in order to achieve specified environmental targets.

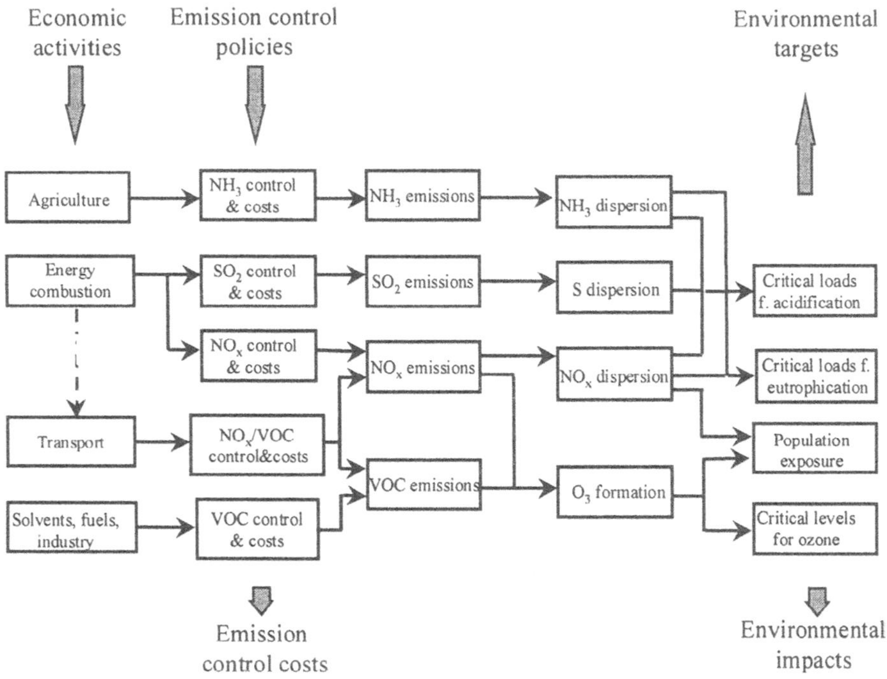

Figure 1: Schematic flowchart of the RAINS model framework

3. The Optimization Problem

The optimization mode of integrated assessment models can be a powerful tool in the search for cost-effective solutions to combat an air pollution problem. The optimization aims at cost-effectiveness by identifying the cost-minimal allocation of resources for emission control in order to meet user-selected environmental air quality targets.

Particularly in the case of tropospheric ozone, a systematic search for cost-effectiveness appears attractive. The facts that several pollutants (NO_x and VOC emissions) are involved, and that important non-linearities between precursor emissions and ozone levels have been recognized, cut the likelihood of 'intuitive' solutions being identified in the scenario analysis mode.

The model distinguishes between a set of I sources of various types of air pollution and a set of J receptor areas for which various air quality targets are assessed. Emissions are analyzed for sets of emitters that are located in a region i which is typically a country. NO_x and VOC emitters are further subdivided into sectors in order to account for sectoral emission controls that reduce either NO_x or VOC or a linear combination of them.

3.1 DECISION VARIABLES

The main decision variables are the annual emissions of the following four pollutants:

n_{is} = annual emissions of NO_x in sector s of country i,
v_{is} = annual emissions of VOC in sector s of country i,
a_i = annual total emissions of NH_3 of country i, and
s_i = annual total emissions of SO_2 of country i.

Additionally, optional variables are considered for limited violations of air quality targets. For such scenarios, variables corresponding to each type of the considered air quality targets are defined for each receptor:

yl_j = violation of ozone exposure targets (surplus if $yl_j < 0$);
ya_j = violation of the acidification targets (surplus if $ya_j < 0$).

Each variable represents a violation of a given environmental target. Violations of targets can be balanced with surpluses at selected other receptors (within the same country).

3.2 AUXILIARY VARIABLES

Auxiliary variables are introduced for total national emissions of NO_x (n_i), total national emissions of VOC (v_i) (summing up all sectoral emission in a country) and the mean effective emissions of NO_x experienced at the j-th receptor (en_{lj}).

3.3 OUTCOME VARIABLES

One outcome variable represents the total costs of emission reductions, summing up the reduction costs for all pollutants. The costs for reducing individual pollutants are described by piece-wise linear functions:

$cs_i(s_i)$	for SO_2,
$ca_i(a_i)$	for NH_3,
$cn_i(n_i)$	for stationary NO_x sources,
$cv_i(v_i)$	for stationary sources of VOC emissions,
$c_i(n_i, v_i)$	for simultaneous NO_x and VOC control at mobile sources.

For each cost function the domain is specified through upper and lower bounds of the arguments, which implicitly define lower and upper bounds for total national emissions. These bounds may be tightened by an optional specification of bounds on total national or sectoral emission, e.g., to reflect upper limits to the emissions related to certain policy commitments.

Another set of outcome variables describe for each receptor the air quality exposure resulting from the given set of emissions.

The RAINS model represents health-relevant ozone exposure by the so-called AOT60, i.e., the excess dose over the WHO guideline value of 60 µg/kg accumulated over a three-months period. The outcome variable $AOT60_j$ is related to the decision variables by

$$AOT60_j = k60_j + \sum_{i=1}^{M}(a60_{ij}v_i + b60_{ij}n_i + c60_{ij}n_i^2) + \alpha60_j\overline{en}_j^2 + \overline{en}_j \sum_{i=1}^{M} d60_{ij}v_i \tag{1}$$

with $k60_j$, $a60_{ij}$, $b60_{ij}$, $c60_{ij}$, $\alpha60_{ij}$ and $d60_{ij}$ as the receptor-specific coefficients of the reduced-form ozone model developed by Heyes and Schöpp [5]. The subset M of the emission sources contains the 12 countries having largest impact on ozone exposure at a given receptor site j.

A similar relation is specified for the vegetation-related excess ozone criterion, i.e., the AOT40:

$$AOT40_j = k40_j + \sum_{i=1}^{M}(a40_{ij}v_i + b40_{ij}n_i + c40_{ij}n_i^2) + \alpha40_j\overline{en}_j^2 + \overline{en}_j \sum_{i=1}^{M} d40_{ij}v_i \tag{2}$$

where $k40_j$, $a40_{ij}$, $b40_{ij}$, $c40_{ij}$, $\alpha40_{ij}$ and $d40_{ij}$ are the site-specific coefficients of the reduced-form ozone model.

For acidification, two types l of outcome variables describe acid deposition ac_{lj} using a piece-wise linear approximation of the critical load function representing pairs of sulfur and nitrogen deposition values that lead to equal protection of the ecosystems in a grid cell j. These variables are related to the decision variables via

$$ac_{lj} = tns_{lj}(\sum_i tn_{ij}n_i + \sum_i ta_{ij}a_i) + tss_{lj}\sum_i ts_{ij}s_i + kn_j + ks_j \tag{3}$$

tn_{ij}, ta_{ij}, ts_{in} are the atmospheric transfer coefficients for NO_x, NH_3 and SO_2 emissions, respectively, kn_j and ks_j the background depositions for sulfur and nitrogen and tns_j and tss_j scaling coefficients to convert sulfur and nitrogen deposition into units of acidity of the critical loads functions. The coefficients tn_{ij}, ta_{ij}, ts_{in} kn_j and ks_j are obtained from the EMEP model; tns_{lj} and tss_j, are extracted from the critical loads database.

From these acid deposition variables (ac_{lj}), the accumulated excess deposition aac_j (i.e., for each grid cell the total acid deposition in excess of the critical loads of all ecosystems) is determined using a piece-wise linear function (PWL_j), which is determined from the critical loads database:

$$aac_j = PWL_j ac(ac_j) \tag{4}$$

For eutrophication, the outcome variable eu_j is linked to the decision variables by

$$eu_j = \sum_i tn_{ij}n_i + \sum_i ta_{ij}a_i + ke_j \tag{5}$$

where ke_j denotes the nitrogen background deposition.

3.4 CONSTRAINTS

Each of the decision variables is implicitly bounded by a corresponding definition of the domain of the cost function

$$n_{is}^{min} \leq n_{is} \leq n_{is}^{max},$$
$$v_{is}^{min} \leq v_{is} \leq v_{is}^{max},$$
$$a_i^{min} \leq a_i \leq a_i^{max},$$
$$s_i^{min} \leq s_i \leq s_i^{max}.$$

The $AOT_{(l=60/40)}$ at each receptor is constrained by user-specified air quality targets

$$AOT_{lj} - y_{lj} \leq AOT_{lj}^{max} \tag{6}$$

and the accumulated excess acidity is constrained by

$$aac_j - ya_i \leq aac_j^{\max} \tag{7}$$

with aac_j^{max} specified by the user.

Optionally, violations of targets can be balanced with surpluses of targets within restricted sets of receptors (J_m), where $m \in M$ is the index of a set of receptors (e.g., all receptors in a country). The balances are represented by the following constraints:

$$\sum_{j}^{Jm} wo_{lmj} y_{lj} \leq tbo_{lm}, \, l = 0 \tag{8}$$

$$\sum_{l} \sum_{j}^{Jm} wo_{lmj} y_{lj} \leq \sum_{l} tbo_{lm} \tag{9}$$

$$\sum_{j}^{Jm} wa_{mj} ya_j \leq tba_m \tag{10}$$

where wo_{lmj} and wa_{mj} are given weighting coefficients, J_m are set of receptors and tbo_{lm}, and tba_{lm} are target balances for the m-th set of receptors.

3.5 GOAL FUNCTION

A composite goal function is used for a single criterion optimization of the non-linear ozone model in order to meet the following goals:

- minimization of total costs of emission reductions,
- minimization of violations of environmental targets.

Therefore, the goal function G is formulated as

$$G = \sum_{i} (ca_i(a_i) + cs_i(s_i) + cn_i(n_i) + cv_i(v_i) + c_i(n_i, v_i)) + penalty \tag{11}$$

The penalty term is defined by

$$penalty = \rho \sum_{l} \sum_{j} y_{lj}^{\xi} + \rho_a \sum_{j} ya_j^{\xi} \tag{12}$$

where ρ and ρ_a are a small positive penalty coefficients.

3.6 SECTORAL COST CURVES AS INPUT TO THE OPTIMIZATION

Inputs to the optimization package include cost curves providing, for the various pollutants under consideration, the least costs for achieving increasingly stringent emission reductions. They are compiled by ranking the available abatement options according to their marginal costs. This methodology produces piece-wise linear curves, consisting typically of about 20-80 segments, depending on the pollutant.

For each of the pollutants and each country, such piece-wise linear curves can be used as input to the optimization. Although the solvers used for this exercise are capable of dealing with piece-wise linear curves, for reasons of increased numerical stability the original piece-wise linear information was smoothed at corners.

4. Implementation

The optimization problem as defined above is a large non-linear model with about 14000 variables (12000 nonlinear), 18000 constraints (3500 nonlinear) and 240 000 Jacobian elements. The problem-specific model generator has been implemented in C++, and the optimization task is presently performed on a SUN Ultra 450 MP workstation.

Two solvers, namely CONOPT [3] and MINOS [6] are used for solving the resulting optimization problem.

5. An Optimized Scenario for Meeting Environmental Interim Targets in Europe

As an example of a cost-optimized scenario presented recently in the course of European negotiations, the optimized reductions for SO_2, NO_x, NH_3 and VOC for meeting the environmental targets listed in Table 1 are provided in Table 2.
It is clear from Table 2 that cost-optimized emission reductions impose differentiated requirements to the individual countries. They are therefore in contrast to the approach adopted in many conventional international environmental agreements, where a notion of equity (expressed, e.g., in terms of equal percentage reductions of emissions compared to a common base year) superseded the cost-effectiveness.

If the overall emission reductions of the optimized scenario were implemented in a uniform way in all 15 EU countries, actual costs would increase by about 10 percent, while the environmental protection (e.g., in terms of protected ecosystems area) would decrease by about 20 percent.

TABLE 1. Summary of the environmental targets for the optimized scenario

Acidification
- Reduction ('gap closure') of accumulated excess acidity everywhere by at least 95 %

Health-related ozone
- Reduction ('gap closure') of the health relevant exposure criterion (AOT60) everywhere by at least 67 %
- Maximum AOT60 of 2.9 ppm.h to be achieved in 4 out of 5 years

Vegetation-related ozone
- Reduction ('gap closure') of the health vegetation exposure criterion (AOT40) everywhere by at least 33 %
- Maximum excess AOT40 of 10 ppm.h, mean over five years

TABLE 2: Emission reductions in percentages (compared to the year 1990) for the optimized (OPT) scenario compared to the Reference case (the present legislation, which is used as the starting point for the optimization). The UNI-OPT columns indicate the optimization results, when emissions are pushed as much as possible towards the EU-15 average reductions while maintaining the environmental targets.

	SO_2			NO_x			VOC			NH_3		
	REF	OPT	UNI-OPT	REF	OPT	UNI-OPT	REF	OPT	UNI-OPT	REF	OPT	UNI-OPT
Austria	-55	-55	-65	-41	-51	-55	-41	-62	-63	-13	-13	-21
Belgium	-38	-81	-81	-41	-68	-63	-47	-74	-70	-1	-38	-29
Denmark	-51	-74	-77	-50	-50	-59	-47	-47	-61	-6	-9	-20
Finland	-50	-50	-73	-41	-41	-58	-47	-47	-65	-23	-23	-23
France	-61	-80	-81	-44	-59	-61	-48	-64	-65	-1	-10	-22
Germany	-88	-91	-92	-53	-60	-63	-63	-69	-71	-25	-48	-41
Greece	12	12	-76	0	-2	-29	-39	-40	-56	-8	-8	-21
Ireland	-61	-82	-77	-28	-42	-56	-59	-59	-64	-1	-4	-11
Italy	-65	-65	-77	-42	-57	-56	-43	-54	-59	-10	-10	-21
Luxembourg	-71	-71	-77	-55	-77	-62	-58	-74	-66	0	0	-5
Netherlands	-63	-74	-74	-42	-52	-57	-51	-69	-67	-42	-55	-55
Portugal	-49	-49	-77	-5	-46	-51	-34	-41	-43	-6	-6	-21
Spain	-64	-65	-77	-23	-29	-50	-36	-36	-51	0	0	-20
Sweden	-44	-44	-53	-41	-46	-55	-42	-52	-61	-21	-21	-21
UK	-74	-86	-88	-58	-58	-61	-49	-61	-64	-10	-20	-22
EU-15	-70	-77	-84	-45	-53	-58	-49	-60	-63	-12	-22	-28

An indication of the savings achieved by the cost-optimization has been derived by another optimization analysis, in which the emission reductions for the individual countries were pushed as much as possible towards the 'uniform' reduction levels, while maintaining the environmental targets (constraints) listed in Table 1.

In practice, the mathematical optimization problem was extended by a 'regularization' term, which puts a (quadratic) penalty on each deviation of an optimized emission reduction level from an exogeneously specified 'target' emission level. The goal function of the optimization problem as presented above is extended by a regularization term

$$\varepsilon \cdot \|z-\check{z}\|^2 \tag{13}$$

where z denotes the vector of the decision variables (emissions relative to 1990) and $\check{z}$ the vector of the uniform 'target' emission levels (relative to 1990).

Results of this optimization analysis are presented in Table 2 under the UNI-OPT columns. With a sufficiently large regularization term (ε=1000), the reduced variability of national emission reductions leads to cost increases of up to 60 percent. As an example, Figure 2 compares the cost-effectiveness of the optimized and non-optimized scenarios in relation to the area of ecosystems protected against acidification

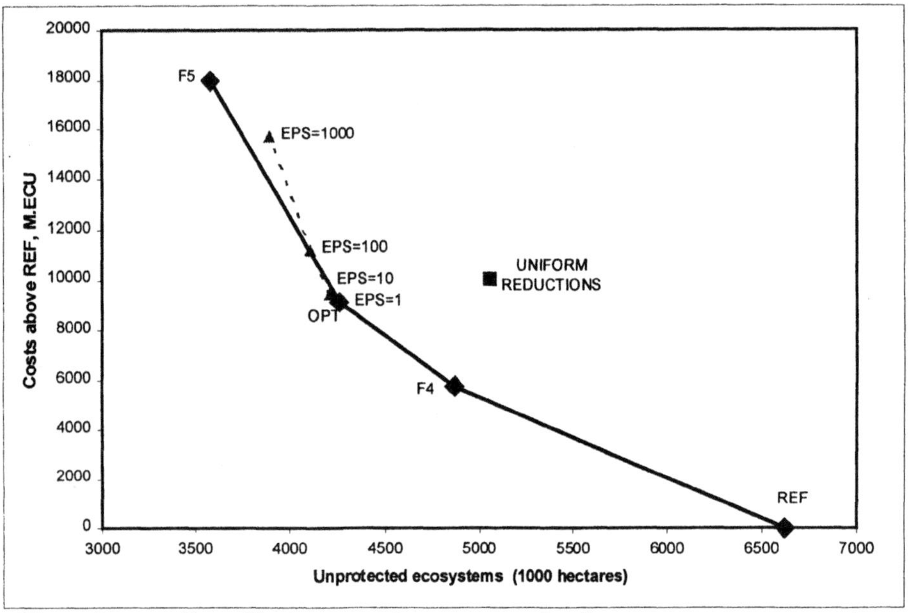

Figure 2. Cost-effectiveness in terms of the ecosystems protection (acidification) of the uniform reduction scenario and the sensitivity runs with different regularization terms ε compared to those of the optimized scenarios

6. Conclusions

Applying an optimization concept to an integrated assessment model for regional air quality management suggests a significant potential for cost-savings compared to traditional approaches for determining emission reduction targets. The optimization takes into account differences in emission control potentials, emission control costs and (non-linear) relations between precursor emission levels and various indicators of air quality in different parts of Europe.

7. References

1. Alcamo, J.; Shaw, R. and Hordijk, L. (eds.) (1990). The RAINS Model of Acidification. *Science and Strategies in Europe.* Kluwer Academic Publishers,. Dordrecht, The Netherlands

2. Barret, K. and Sandnes, H., (1996). Transboundary Acidifying Air Pollution calculated transport and exchange across Europe, 1985-1995. In: Barret, K. and Berge, E. (eds.)*Transboundary Air Pollution in Europe*. MSC-W Status Report 1996, Meteorological Sythesizing Centre - West, Norwegian Meteorological Institute, Oslo, Norway.

3. Drud, A. (1996). *CONOPT:* A system for large scale nonlinear optimization. ARKI Consulting and Development A/S, Bagsvaerd, Denmark

4. EEA, (1996). *Joint EMEP/CORINAIR'90 Atmospheric Emission Inventory Guidebook, First Edition.* Vol.1-2. European Environmental Agency, Copenhagen, Denmark.

5. Heyes C. and Schöpp W. (1997). A 'Reduced-Form' Model to Predict Long-Term Ozone Concentrations in Europe, International Institute for Applied Systems Analysis, Laxenburg, Austria

6. Murtagh, B. and Saunders, M., (19870. MINOS 5.1 User's guide. Technical Report SOL 83-20R, Stanford University, Standvord, CA, USA.

7. Posch, M.; Hettelingh, J.-P.; de Smet, P.A.M. and Downing, R.J. (eds.) (1997) Calculation and mapping of critical thresholds in Europe. Status Report 1997, Coordination Center for Effects, RIVM, Bilthoven, Netherlands.

8. Simpson, D. (1993). Photochemical model calculations over Europe for two extended summer periods : 1985 and 1989. Model calculations and comparison with observations*Atmos. Environ.*, **27A**, No. 6, pp. 921-943.

SPATIAL REFINEMENT OF REGIONAL EXPOSURE ASSESSMENT

VOLKER BERDING, FRANK KOORMANN, STEFAN SCHWARTZ, JAN-OLIVER WAGNER, MICHAEL MATTHIES
Institute of Environmental Systems Research,
University of Osnabrück
49069 Osnabrück
Germany

Abstract

Using the *European Union System for the Evaluation of Substances* (EUSES) and the *Geography-referenced Regional Exposure Assessment Tool for European Rivers* (GREAT-ER) regional exposure assessments were performed for the polycyclic musk fragrance HHCB (1,3,4,6,7,8-hexahydro-4,6,6,7,8,8-hexamethyl-cyclopenta-[g]-2-benzopyrane; trade name: e.g. Galaxolide®). Starting with a generic standard region a spatial refinement was carried out for the German River Ruhr region. The refinement was realised in different scenarios by successively replacing EUSES default parameters with realistic regional values and then applying the selected region to GREAT-ER. The results were compared to monitoring data from the region of North Rhine-Westphalia (River Ruhr). It was shown that EUSES estimates the median of the measured values very well in every scenario. Spatial refinement leads to lower concentrations. Even underestimations are possible if realistic regional parameters are inserted and a ready biodegradability is assumed. The lowest deviations to measured values are the average concentrations calculated by GREAT-ER, however, much more data are needed to perform a reasonable regional assessment. Furthermore, assuming the same region, the predicted concentrations of EUSES and GREAT-ER do not differ by more than a factor of 5. In addition, GREAT-ER delivers realistic regional information with visualised concentration profiles and maps.

1. Introduction

In 1996 the European Commission published the Technical Guidance Document (TGD, [1]), which enables the assessment of the risk posed by new notified and existing chemicals to humans and the environment. Three tiers are proposed for risk assessments: tier 1 as an initial (screening) stage, tier 2 as an intermediate (refined) stage and tier 3 as a comprehensive stage. The models and procedures laid down in the TGD have

J.B.H.J. Linders (ed.), Modelling of Environmental Chemical Exposure and Risk, 205–222.

been made available in software form by the European Union System for the Evaluation of Substances (EUSES, [2]) which is primarily intended to be used as a screening tool (tier 1). EUSES makes default values for most of the parameters available; physico-chemical substance properties and the tonnage are required as input parameters. The default values claim to represent a hypothetical "European average region", the so-called standard region [1].

The Geography-referenced Regional Exposure Assessment Tool for European Rivers (GREAT-ER) is a support tool for environmental risk assessment and river basin management, which focuses on a more refined stage of risk assessment (tier 2) [3]. It uses a Geographic Information System (GIS) to produce a visualisation of predicted chemical concentrations along a river [4]. Additionally, it is possible to produce profiles of Predicted Environmental Concentrations (PECs) throughout the studied catchment. Such profiles illustrate chemical fate from a river's headwaters down to its mouth, and can be used to compare model predictions with monitoring data. Further, geo-referenced model results can be aggregated to obtain a spatially averaged PEC, which is representative of the river basin under investigation [5].

The reference substance for this investigation is the synthetic polycyclic musk fragrance HHCB (1,3,4,6,7,8-hexahydro-4,6,6,7,8,8-hexamethyl-cyclopenta-[g]-2-benzopyrane; CAS-No. 1222-05-5; trade name: e.g. Galaxolide®). Synthetic musk fragrances are an essential ingredient in numerous perfumes, cosmetics and cosmetic care products, soaps, detergents, and other cleaning agents [6]. Since the early 1990s not only nitro musk compounds [7], but also polycyclic musk fragrances have been detected in rivers and the sea, fish, human adipose tissue and human milk [8-13]. Some of the polycyclic compounds are the prevailing musk fragrances in Western Europe with concentrations in aquatic biota that exceed those of the nitro musks by up to three orders of magnitude [14]. Seven single compounds are involved, of which HHCB occurs with the highest concentrations in the environment. As early as the 1970s scientists became aware of this substance class because of the neurotoxic effects of the representative compound Versalide® [15]. HHCB shows ubiquitous occurrence in the aquatic environment and a high bioaccumulation potential.

Despite its widespread usage, a consumption rate of up to 2,400 t/a HHCB in Europe [16], this chemical has not yet been investigated sufficiently [10]. In addition to data concerning the effects of these substances, information on their environmental fate and distribution is also necessary. Aquatic fate assessments with EUSES were already carried out by Schwartz *et al.* [17, 18]. However, a refined assessment including geographical information is still lacking.

The idea of this study is to start with the EUSES standard region and to fit it in several scenarios to a specific region and finally to use GREAT-ER for geographical-referenced modelling. Our purpose is to determine how spatial refinement can improve the results of an environmental exposure assessment. The deviations of predicted environmental concentrations from measured values obtained by monitoring studies are of particular interest. A further objective is to determine, by consideration of various scenarios, the impact of different model assumptions on the results and to point out possibilities to reduce over- and underestimations. The state of North Rhine-Westphalia

(Germany) and the river Ruhr region which lies in North Rhine-Westphalia (Figure 1) were chosen for the comparison of measured with predicted concentrations. The river Ruhr is a relatively significant river with a length of 217 km, a catchment area of 4,488 km^2 and an average annual mouth outflow of more than two billion m^3 [19]. It is of major importance to the drinking water supply for the surrounding conurbation.

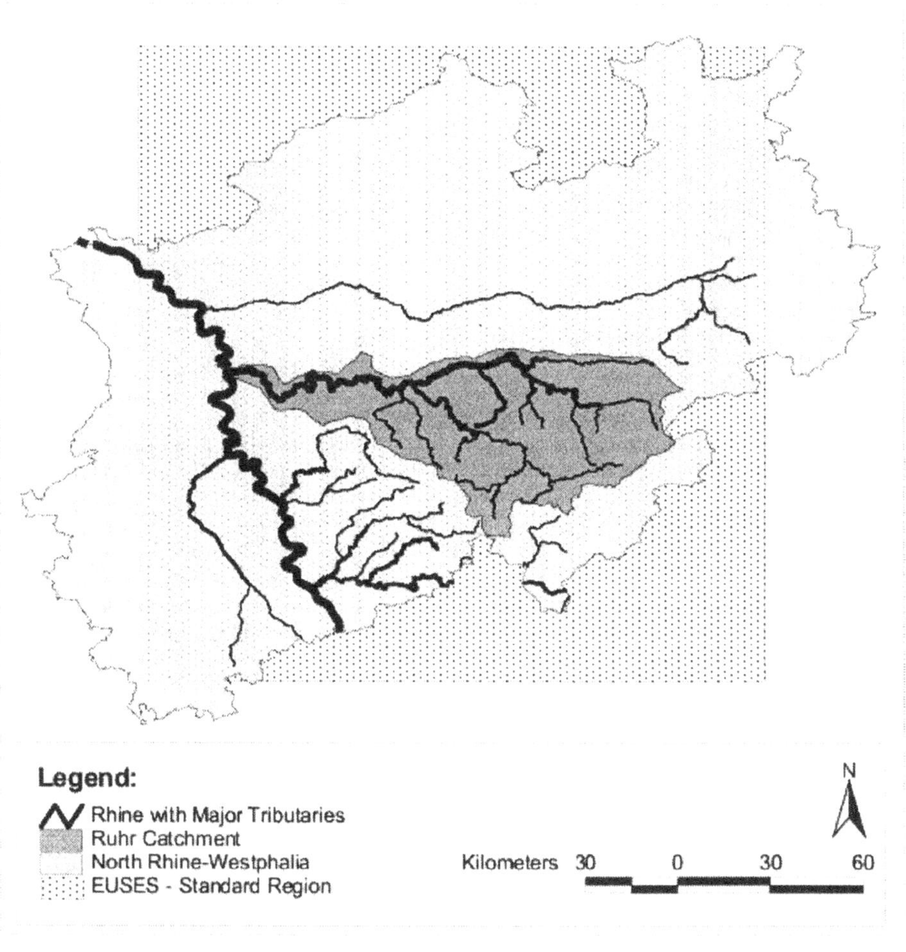

Figure 1. Regions considered for the spatial refinement

For the spatial refinement a stepwise approach is pursued: Firstly, a scenario analysis is performed with EUSES to discover how spatial refinement can ameliorate the predicted concentrations. In a subsequent step the data of the Ruhr catchment are entered in GREAT-ER, and concentrations along the river are calculated. The aggregated results are compared with monitoring data and the results of the EUSES scenario calculations.

2. Models and Databases

In the following the structure, outputs and data requirements of EUSES and GREAT-ER are described briefly. The method of performing a spatial refinement is also presented. Finally, the monitoring data used for the comparison of calculated and measured water concentrations are explained.

2.1 EUSES

2.1.1 Model structure and output

EUSES is a complex system of interconnected models and modules [20]. A release estimation module estimates direct emissions to air, water, waste water and soil. The waste water emissions represent the input for the sewage treatment (STP) model Simple Treat. Simple Treat is a Level III-model that calculates the fate of a substance in an STP and estimates indirect emissions to air, surface water and agricultural soil. These and the estimated direct emissions are used in several local models to predict concentrations in the vicinity of a point source and in a regional model that estimates environmental background concentrations. According to this work's purpose to compare regional aquatic PECs with monitoring data, EUSES is mainly explained with its regional fate model Simple Box. Simple Box is a 6-box Level III model with the compartments air, water, sediment, agricultural soil, industrial/urban soil and natural soil. It contains two nested scales: The outer scale (continental) represents the European Union whilst the inner scale (regional) is a fictive average European region. Simple Box is implemented to calculate regional and continental background concentrations (PECs) for air, water, sediment, three soils and pore water of agricultural soil.

2.1.2 Data requirement of EUSES

By default, EUSES only needs a minimal set of data to carry out an exposure assessment. These are physico-chemical parameters, the biodegradation class and production and usage data of the evaluated substance. Further information is required for a spatial refinement: Firstly, regional parameters are needed to fit the standard region to realistic properties. Then, more exact substance data are necessary to replace estimated values by realistic ones.

Regional parameters. The default parameters can be replaced by data from a specific region, like North Rhine-Westphalia or the Ruhr region. Regional parameters, which differ from the standard are summarised in Table 1. Obviously, the standard region is the largest one, North Rhine-Westphalia is slightly smaller and the Ruhr catchment is the smallest. But spatial refinement does not only mean scaling down the area, but other relevant parameters must also be changed, e.g. precipitation rate, number of inhabitants or the fraction connected to sewer systems.

TABLE 1 Characteristics of the investigated regions used in EUSES

Parameter	Standard [2]	North Rhine-Westphalia [21]	River Ruhr catchment [19]
Area of region [km²]	44,000	34,400	4,488
Average annual precipitation [mm/a]	700	679	1,056
Sewage treatment: BOD50 [g/d]	54	60	60
Sewage treatment: Mode of aeration	surface	bubble	bubble
No. of inhabitants discharging into one STP	10,000	17,226	22,804
Fraction connected to STP [-]	0.70	0.92	0.95
Depth of water [m]	3	3	2
Fraction of water area in region [-]	0.030	0.018	0.018
Fraction of water flow from continental scale to the region [-] (= Area of region / Area of EU · Depth of water)	0.034	0.029	0.002
No. of inhabitants in region	$2.00 \cdot 10^7$	$1.78 \cdot 10^7$	$2.21 \cdot 10^6$

Substance data. A minimum of four physico-chemical parameters is necessary to run an EUSES calculation (Table 2). Estimated partition coefficients and bioconcentration factors may be replaced by more reliable (e.g. measured) values on demand. The physico-chemical data applied in this work are experimentally determined and listed in Table 2. Reasonable data on aquatic degradation have not yet been published, thus, for the purpose of a worst-case estimation, no degradation is assumed. However, Van de Plassche and Balk [16] allude to a possible aquatic biodegradation. To allow for this eventuality, in one scenario HHCB was classified as *readily biodegradable.* EUSES estimates the required degradation rates based on this biodegradability class: A half-life of 15 days for bulk surface water and for sewage treatment plants a half-life of 0.7 hours is assumed.

EUSES estimates emissions on both the continental and regional scale from a given continental tonnage using emission tables for a certain *use category.* Since HHCB is mainly applied as a cleaning/washing additive, this use category is selected in EUSES. Alternatives, e.g. *cosmetics* or *odour agents*, only have a negligible impact on the results. Determining the correct continental tonnage is more difficult since only a few production and consumption rates of HHCB have been published to date: Ohloff [6] states an annual production volume of about 1,000 t world-wide, whereas Van de-Plassche and Balk [16] quote a consumption of 2,400 t (1992) and 1,482 t (1995) in Europe, based on data provided by the producers. The substance is emitted into waste water during use, minus a negligible loss due to volatilisation (fragrance!). Thereafter, it reaches the surface waters via municipal sewage treatment plants, or agricultural fields via sludge application.

In this investigation the regional tonnage is estimated in two ways: Firstly, it is estimated by EUSES, whereby 10% of the continental tonnage is taken as a regional tonnage, i.e. a rate of 240 t/a is given for 1992. Secondly, the tonnage is calculated on the basis of the average per-capita consumption of 18 mg/d [16] and the number of inhabitants in the region. Thus, a consumption rate of 117 t/a is obtained for the region of North Rhine-Westphalia and 14.5 t/a (1992) for the River Ruhr catchment, respectively.

*TABLE 2. Physico-chemical data [16]. Parameters marked with * were estimated by EUSES.*

Parameter	Value	Unit
Molar mass	258.4	g/mol
Vapour pressure	0.0727	Pa, 25°C
Henry's law constant*	11.3	Pa·m³/mol
Log K_{OW}	5.9	-
Log K_{OC}*	4.86	l/kg
Water solubility	1.75	mg/l, 25°C

2.1.3 *Spatial refinement with EUSES (scenario analysis)*

The majority of estimated and default parameters must be changed if a real region needs to be represented. Besides using the standard scenario of EUSES, three different emission and distribution scenarios were calculated (Table 3). The scenario *NRW* considers a regional tonnage based on per-capita consumption and the number of inhabitants. The partition coefficients calculated by EUSES are replaced by the measured values in Table 2. The scenario *Ruhr 1* reduces the spatial scale to that of the River Ruhr catchment, i.e. data specifying the region (e.g. area of region, number of inhabitants, fraction of population connected to sewage treatment plants) were aligned where possible. In scenario *Ruhr 2* ready biodegradability of HHCB is additionally assumed. The differences in the resulting concentrations and their comparison to measured values enable us to determine to what extent an adjustment of the standard environment towards a selected region and an alteration of model assumptions influence the predicted water concentrations.

TABLE 3. Survey of the EUSES scenarios

Scenario name	Region	EU Production volume [t/a]	Regional tonnage [t/a]	Measured partition coefficients	Biodegradability	Description / Characteristics
Standard	Standard	2,400	240	no	no	Input of the minimal parameter set.
NRW	NRW	2,400	117	yes	no	Regional tonnage is based on per-capita consumption and the number of inhabitants.
Ruhr 1	Ruhr	2,400	14.5	yes	no	As NRW, but regional-specific parameters for the river Ruhr catchment are used.
Ruhr 2	Ruhr	2,400	14.5	yes	ready	As Ruhr 1, but ready biodegradability is assumed.

2.2 GREAT-ER

The aim of the GREAT-ER-project was to develop a tier 2 simulation tool for the water compartment in the framework of the environmental risk assessment described in the TGD. In addition to the development of the model system and the ArcView-based user interface, the project included hydrologic modelling and data integration for the sample catchments in the UK and Italy, and extensive monitoring programmes [22].

2.2.1 Model structure

The simulation system of GREAT-ER can be separated into two sections: 1) the waste water pathway with its modules emission, sewer and treatment plant, and 2) the river. These are connected and form a chain of models (Figure 2). To deal with reality (e.g. changing hydrologic situations or confluences) river modules can be linked to a network (Figure 3).

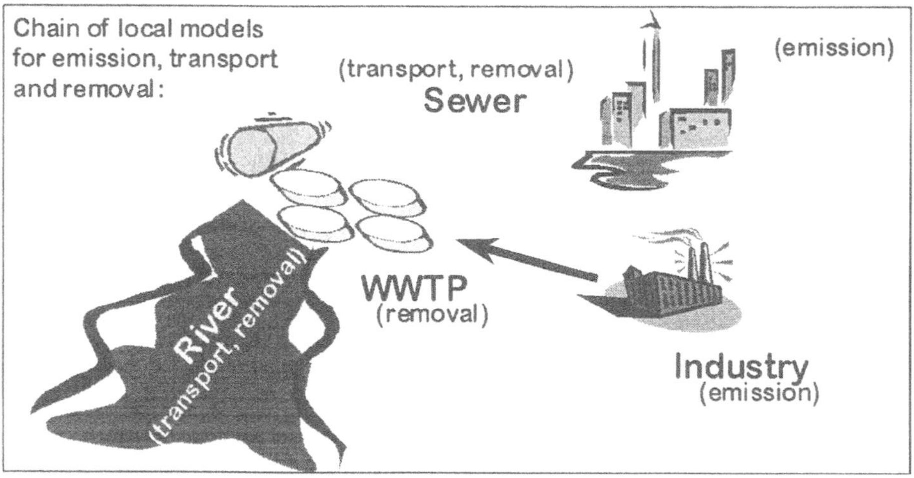

Figure 2. Chain of models

Figure 3. Network of chains/river modules

Depending on the availability of data, several complexity modes can be selected for the different submodels. The complexity ranges from simple conceptual models to highly resolved systems covering several physical models, and hence the information needed varies from easily available parameters to comprehensive data sets.
The models are steady-state deterministic. On top of this core a stochastic approach is added, applying a Monte-Carlo simulation to deal with variability and uncertainty. Hence, the parameters concerned also have to provide this information and the model results are distributions of concentrations in time.

Waste water pathway. The waste water pathway module is separated into three models:

- The **emission module** estimates the input into the waste water system based on the per capita consumption and the number of people connected to the system. This is the core information, which must be available for each application.
- The **sewer module** can be used to consider loss of the substance under investigation on its way from the consumption/use to the sewage treatment plant by a percentage elimination rate. However, it can be switched off.
- The **sewage treatment plant module** represents the complexity mode approach more clearly: In general a primary settler and an activated sludge or trickling filter process can be selected. The minimum data required is the mean flow of the treatment plant and a substance-specific percentage elimination rate for the considered processing steps. The mass flux from the sewer system is considered as input into the treatment plant. Additionally, a second mass flux can be specified to deal with industrial discharges.

  If the required data is available for primary settler and activated sludge processing, more complex models can be selected. The mass balancing is based on a Level III-model and primary settler (4-box) and activated sludge (6-box) models use the Simple Treat 3.0 approach [23]. It has to be noted that most of the parameters of the Simple Treat approach must be provided for each single treatment plant. Boeije [24] also developed a complying trickling filter model which is currently not part of the GREAT-ER model system.

The final result of the waste water pathway model chain is a concentration distribution in the effluent of the treatment plant. The distribution for each separate plant is considered as input into the related river stretch.

River. Unlike the waste water pathway, the river models are not constrained to a chain. As shown in Figure 3 river modules can be linked to a chain to deal with changes in the hydrologic conditions or, more generally, to a network to consider confluences and bifurcations.

Altogether the three models representing the complexity modes are based on a first-order decay:

$$c(x) = c_0 \cdot \exp(-k\, t_x) \tag{1}$$

where

$c(x)$	=	concentration at x
c_0	=	initial concentration in the river stretch
k	=	first-order elimination rate
t_x	=	travel time to x (calculation based on length and flow velocity)

The models differ in the degree of detail of calculating the first-order elimination rate.

- Mode 1 considers an aggregated overall first-order rate which can be obtained from literature on field studies or 'simple' experiments.
- The second mode covers general degradation and also loss by sedimentation (sorbed fraction) and volatilisation (solved fraction). The final decay rate is calculated by the sum of the separate rates. Considering the fractioning process increases the amount of parameters required (both geography-referenced and substance related) significantly.
- The third complexity mode considers elimination (black boxes given by their elimination rates in modes 1 and 2) as individual processes represented by detailed mechanistic models. These cover the processes of hydrolysis, photodegradation, biodegradation, sedimentation and volatilisation. Depending on the physico-chemical properties of the substance under investigation separate processes can be switched on/off to control the number of parameters needed.

2.2.2 *Model output*

A simulation run results in separate concentration distributions (over time) per stretch and per treatment plant. From this large set of data information can be obtained comfortably in different ways. Since the GREAT-ER user interface is GIS-based, at first the calculated concentration mean values are presented on a colour coded map to give an easy overview of the study area. Since concentration distributions are calculated any percentile can be derived from the data for visualisation. For a more detailed analysis of model results concentration profiles can be generated for selected streams. Finally, two concepts are put under scientific investigation to derive regionalised information from the highly detailed data [5]:

- $PEC_{initial}$, the temporal mean of all concentrations in the river network directly after discharge
- $PEC_{catchment}$, the weighted spatial and temporal mean of all (polluted) stretches in a catchment.

2.2.3 Data requirements of GREAT-ER

Depending on the aspired complexity mode for the simulation (see Section 2.2.1), the data requirements for GREAT-ER can vary enormously. The most important factor to be considered is that the attributes of the geography-referenced objects (e.g. sewage treatment plants) must be available for each single object. Since this is only possible with in-depth investigation and exhaustive data exploration, the exercise presented in this paper is based on the simple model modes of GREAT-ER and hence on the minimum required data set.

2.2.4 River Network / Hydrology

Considering in-stream removal as a process of simple first-order decay three attributes are required per stretch: length, flow and flow velocity. The length is given by the underlying data set of the UBA (Gewässer-GIS Deutschland/Umweltbundesamt). The flow was estimated applying a simple regression-based hydrologic model: Dyck [25] and many others mention a relationship between the accumulated length and the flow:

$$Q = a \cdot L_{acc}^{\ b} \qquad (2)$$

Where

Q	=	flow [m³/s]
L_{acc}	=	accumulated length [m], the sum of all stretches' lengths upstream from a given stretch
a, b	=	regression parameters.

The regression equation was fitted based on the Ruhr catchment data taken from the UBA river network (length) and the German Hydrologic Yearbook [26] (flow data based on long-term statistics). The flow velocity was derived applying a second hydrologic model by Round et al. [27]:

$$V = 10^{-0.583} \cdot MQ^{0.283} \cdot (Q/MQ)^{0.495} \qquad (3)$$

where

V	=	flow velocity [m/s]
MQ	=	mean flow [m^3/s]
Q	=	flow [m^3/s].

The hydrology of the Ruhr catchment is shown in Figure 4.

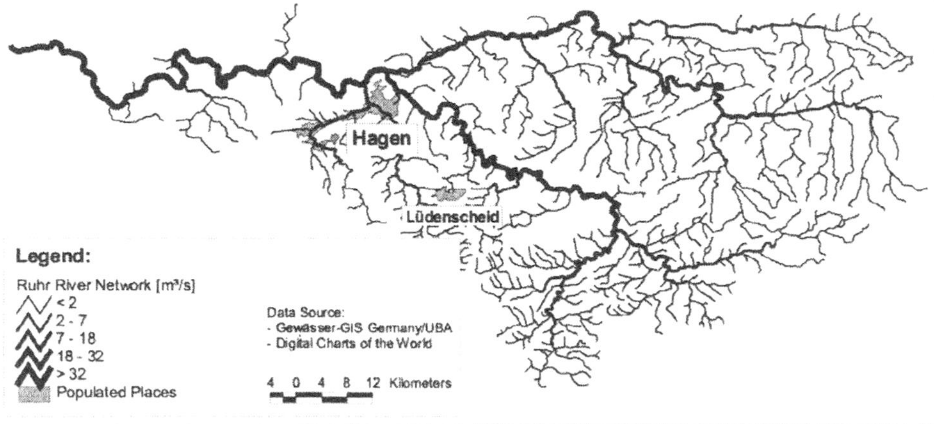

Figure 4 Hydrology of the Ruhr catchment

2.2.5 Emission / Sewage Treatment Plants

GREAT-ER considers geography distributed emissions from point sources. The information regarding the treatment plants was provided by the Ruhrverband (Institution for River Ruhr Water Management) [28]. The data set covers a set of 96 treatment plants in the Ruhr catchment run by the Ruhrverband (see Figure 5). There are further (especially) industrial treatment plants, but since HHCB is usually released into the sewage by domestic use the Ruhrverband treatment plants (serving about 2.12 million inhabitants) are sufficient. Beside the location (obviously necessary) the data set provides the number of connected people and the mean daily flow based on data from 1998.

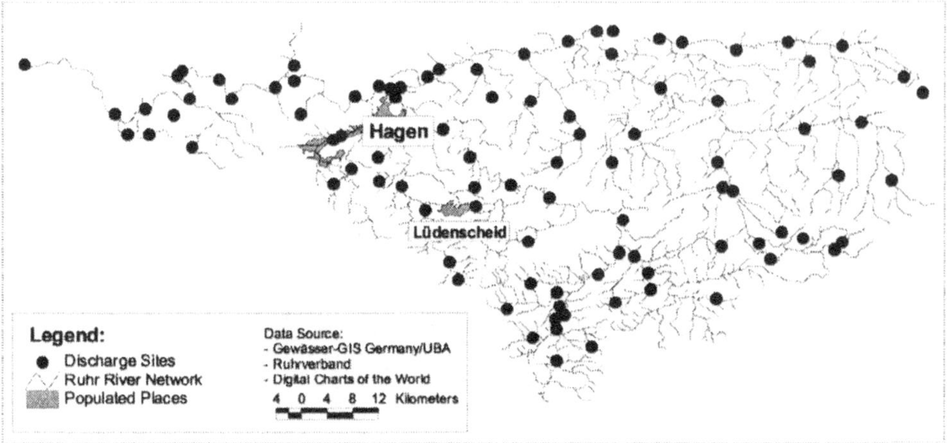

Figure 5 Sewage Treatment Plants of the Ruhrverband in the Ruhr catchment

2.2.6 *Substance data*

In the configuration used GREAT-ER needs even less substance information than EUSES: only degradation data (for water and sewage treatment plant) and per-capita use are required. The degradation rates, which are the values calculated by EUSES, are shown in Table 4.

TABLE 4 Substance data for GREAT-ER

Parameter	Value	Unit
Treatment Plant Elimination	94	%
In-stream removal rate	$1.93 \cdot 10^{-3}$	1/h

2.3 MONITORING DATA

Measured concentrations from the river Ruhr [9, 10] were used to compare the model's results with monitoring data. The measurement of 30 samples at the beginning of 1994 along a stretch of 160 km was carried out downstream from three sewage treatment plant discharges as well as at locations without the direct influence of emitters. Sampling was carried out daily over the period of a week.

The HHCB content in the Ruhr is comparatively high, with a mean of 0.37 µg/l. In the river Elbe and the river Rhine average concentrations of 0.12 µg/l [13] and 0.07 µg/l [16], respectively, were measured.

The average monitoring data from the river Ruhr are viewed as regional concentrations from the region of North Rhine-Westphalia. They are compared to the simulation results of the EUSES scenario calculations and the average catchment concentration of GREAT-ER simulations.

3. Results and discussion

EUSES and GREAT-ER estimate concentrations in surface water. The aquatic PECs estimated by EUSES represent regional background concentrations in surface water. The median of the measured values is chosen for a comparison of modelled results with monitoring data. The comparative GREAT-ER-value is the $PEC_{catchment}$, which is the weighted spatial and temporal mean of all (polluted) stretches in a catchment, representing an average concentration of the Ruhr catchment.
When comparing calculated with measured HHCB surface water concentrations, one must bear in mind that the range of measurements is quite narrow (between approx. 0.1 and 1 µg/l). Since the measurements were taken early in the year they may only represent an extract of the annual concentration range. However, calculated concentrations in surface water show relatively small deviations to measured concentrations (Figure 6).

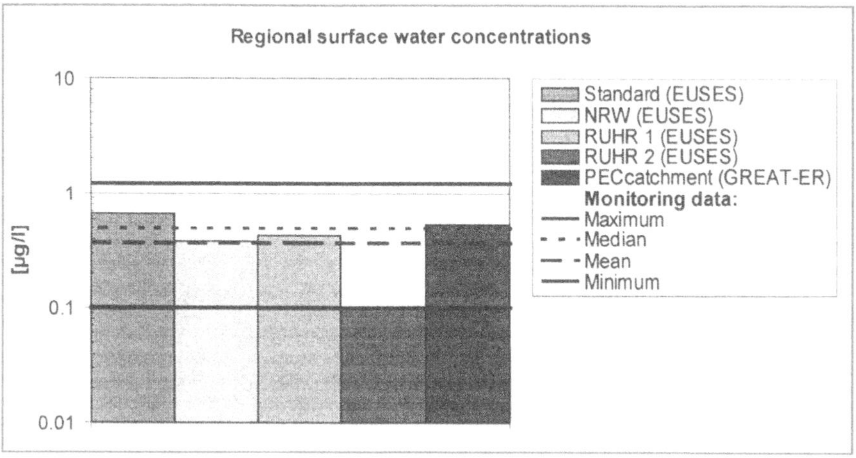

Figure 6. Comparison of calculated surface water concentrations with monitoring data

On a regional scale, the first three scenarios are close to the measured median, over- and underestimating it only slightly. The median of the concentrations measured in the Ruhr represents a mixture of local and regional concentrations, and can therefore be viewed as an increased regional contamination. This is also confirmed by the fact that lower contamination was measured in rivers such as the Elbe and the Rhine. On this basis EUSES delivers rather realistic estimations on a regional scale.
In the sense of the TGD, the measured values are rather to be classified as regional, since local concentrations are always defined by the direct vicinity of an emitter. The variation in the regional scale results of the individual scenarios does not exceed one order of magnitude. It is remarkable that the concentration of the *Ruhr 1* scenario nearly equals that of the *NRW* scenario. This shows that scaling down the environment from the North Rhine-Westphalian region to the Ruhr catchment only influences the resulting regional concentrations negligibly. The predicted concentration of the *Ruhr 1* scenario

corresponds to the median of the measured values very well, but the assumption of ready biodegradation (*Ruhr 2* scenario) leads to distinct underestimations. There may be several reasons for this: Firstly, the supposed degradation rate in water could be too high, which could be reasoned by the fact that the *Ruhr 1* scenario delivers much better estimations. Secondly, the emissions to surface water could be underestimated. Thirdly, the underlying measured values for this investigation may perhaps not be representative and too high since they were all taken in the same winter month (February 1994) along the river Ruhr. The mean air temperature in February 1994 was about 2°C [29]. (For a comparison with temporal variability see [17]). And another reason could be that EUSES generally underestimates regional concentrations and delivers only conservative predictions if the used estimations and defaults are conservative.
The latter presumption is strengthened by the GREAT-ER results: The $PEC_{catchment}$ predicted by GREAT-ER equals the measured median, although it uses the same high degradation rates as the underestimating *Ruhr 2* scenario. The reason for this is that the average concentration carried out by GREAT-ER is a weighted accumulation of geography-referenced simulation results while EUSES assumes one homogeneous box. Besides, the water volume assumed in EUSES does not need to be the same as in GREAT-ER. In the *Ruhr* scenarios and the *NRW* scenario the same area fraction and depth for water are considered while the GREAT-ER calculation is based on realistic river data. Another important functionality in GREAT-ER is that colour coded maps to visualise spatial distribution of estimated concentrations (Figure 7) can be produced.

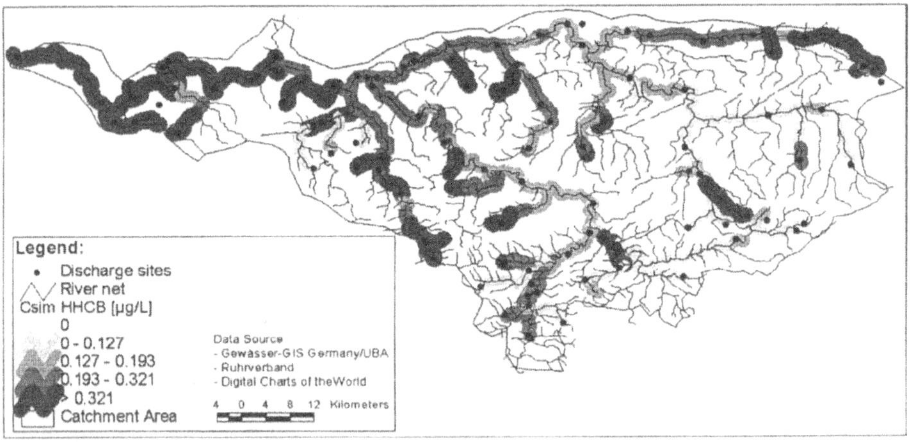

Figure 7. Colour coded map of HHCB concentrations in the Ruhr catchment. C_{sim} is the average concentration in the appropriate river module.

A slight improvement of the EUSES model results can be gained by replacing the standard region by the NRW-specific regional data and by the use of measured partition coefficients. More precise emission rates lead to a considerable improvement of the model results. However, the values used are still afflicted by uncertainty: The consumption quantities are based on production figures provided by the fragrance industry.

These figures do not contain information on import and export trade outside the EU. Nor do they include all manufacturers. Furthermore, it is not known whether the reduction in consumption represents a general trend or fluctuations over a number of years [16]. This could be a significant source of uncertainty regarding statements on the comparison of measured vs. predicted data. But the figures can be taken as a best estimate, and it is expected that qualitative statements (i.e. over- or underestimations) do not change, because in our scenarios the emission rate shows a linear impact on the predictions and it is expected that variations in emissions do not exceed one order of magnitude.
The classification of the substance as readily degradable leads, of course, to noticeably lower environmental concentrations. Suspended matter can influence the distribution behaviour of substances considerably. Winkler *et al.* [13] ascertained a dissolved fraction of 92% for HHCB in the Elbe with a median partition coefficient between water and suspended matter of 4,500 l/kg. The values generated by EUSES are 88% and 1,900 l/kg, respectively. According to this, adsorption to suspended matter has a negligible influence on the result.

4. Conclusions

Both EUSES and GREAT-ER can be applied to evaluate the fate of the fragrance HHCB in the aquatic environment. In the EUSES standard scenario the predicted concentration is slightly higher than the measured median. Hence, the estimations are conservative with the default settings. But EUSES does not necessarily deliver conservative estimations since the NRW scenario and the two Ruhr scenarios underestimate the measured median. The use of a high biodegradability leads to the highest underestimation, but the value is still within the range of the measured values. However, the deviations are low for regional concentrations in the surface water. The variations of water concentrations between the individual scenarios are smaller than the difference between the minimum and maximum of the measured values. From the comparison of the concentrations of the investigated regions it can be concluded that an exact selection of representative environmental segments, including an adjustment of the emission rates to the selected region, leads to more realistic modelling of the contamination of the rivers. This requires closer scrutiny of the monitoring data and their spatial classification.
It is shown that the use of a geography-referenced model leads to good estimations of the median of the measured concentrations while on the same basis (*Ruhr 2* scenario) the compartment model underestimates the median. Additionally, only a geography-referenced model can show hot spots and can produce maps with visualised information. Nevertheless, the results of EUSES and GREAT-ER do not differ more than a factor of 5 if the same data base is applied. In view of this fact one must weigh if the information gain is sufficiently high to justify the large amount of additionally required data.
In this paper we dealt with potential uncertainties in an aquatic fate assessment using different scenarios and different types of models. The uncertainty of substance parame-

ters were not analysed and the sewage treatment process of HHCB was not considered in detail; the necessary database for such a study is not yet available.

5. Acknowledgement

Funding by the German Federal Environmental Agency, Berlin is gratefully acknowledged (R+D project FKZ 206 01 075). We also thank the Ruhrverband, Essen for its support.
GREAT-ER was sponsored by the Environmental Risk Assessment Steering Committee (ERASM) of the Association Internationale de la Savonnerie, de la Détergents et des Produits d'Entretien (AISE) and the Comité Européen des Agents de Surface et Intermédiaires Organiques (CESIO) and the UK Environment Agency, and was managed by ECETOC.

6. References

1. EC (1996). Technical Guidance Document in Support of the Commission Directive 93/67/EEC on Risk Assessment for New Notified Substances and the Commission Regulation (EC) 1488/94 on Risk Assessment for Existing Substances, Parts I–IV. Office for Official Publications of the European Communities, Luxembourg

2. EC (1996). EUSES – the European Union System for the Evaluation of Substances. Institute of Public Health and the Environment (RIVM), The Netherlands. Available from European Chemicals Bureau, Ispra

3. ECETOC (1999). GREAT-ER User Manual. Special Report No. 16. ECETOC, Brussels

4. Matthies, M.; J.-O. Wagner; F. Koormann (1997). Combination of Regional Exposure Models for European Rivers with GIS Information. In: *Proceedings of ECO-INFORMA '97* (Eds. K. Alef *et al.*), Vol. 12, 523–529, Eco-Informa Press, Bayreuth

5. Boeije, G.; J.-O. Wagner; F. Koormann; P. Vanrolleghem; D. Schowanek; T. Feijtel (1999). New PEC Definitions for River Basins applicable to GIS-based Environmental Exposure Assessment (Contribution to GREAT-ER #8), *Chemosphere* (accepted)

6. Ohloff, G. (1990). Riechstoffe und Geruchssinn: Die molekulare Welt der Düfte. Springer, Heidelberg

7. Rimkus, G.; H. Brunn (1996). Synthetische Moschusduftstoffe – Anwendung, Anreicherung in der Umwelt und Toxikologie; *Teil 1: Herstellung, Anwendung, Vorkommen in Lebensmitteln, Aufnahme durch den Menschen. Ernährungs-Umschau* **43**, 442–449

8. Bester, K.; H. Hühnerfuss; W. Lange; G. Rimkus; N. Theobald (1998). Results of non target screening of lipophilic organic pollutants in the German Bight II: Polycyclic Musk Fragrances. Water Research **32**, 1857–1863

9. Eschke, H.D.; J. Traud; H.J. Dibowski (1994). Untersuchungen zum Vorkommen polycyclischer Moschus-Duftstoffe in verschiedenen Umweltkompartimenten – Nachweis und Analytik mit GC/MS in Oberflächen-, Abwässern und Fischen (1. Mitteilung). *Z. Umweltchem. Ökotox.* **6**, 183–189

10. Eschke, H.D.; J. Traud; H.J. Dibowski (1995). Untersuchungen zum Vorkommen polycyclischer Moschus-Duftstoffe in verschiedenen Umweltkompartimenten – Befunde in Oberflächen-, Abwässern und Fischen sowie in Waschmitteln und Kosmetika (2. Mitteilung).*Z. Umweltchem. Ökotox.* **7**, 131–138

11. Eschke, H.D.; J. Traud; H.J. Dibowski (1995). Nachweis und Quantifizierung von polycyclischen Moschus-Duftstoffen mittels Ion-Trap GC/MS/MS in Humanfett und Muttermilch.*Dt. Lebensmittel-Rundschau* **12**, 375–379

12. Rimkus, G.; M. Wolf (1996). Polycyclic musk fragrances in human adipose tissue and milk.*Chemosphere* **33**, 2033–2043

13. Winkler, M.; G. Kopf; C. Hauptvogel; T. Neu (1998). Fate of artificial musk fragrances associated with suspended particulate matter (SPM) from the River Elbe (Germany) in comparison to other organic contaminants. *Chemosphere* **37**, 1139–1158

14. Gatermann, R.; J. Hellou; H. Hühnerfuss; G. Rimkus; V. Zitko (1999). Polycyclic and Nitro Musks in the Environment: A comparison between Canadian and European Aquatic Biota.*Chemosphere* **38**, 3431–3441

15. Spencer, P.S.; A.B. Sterman; D.S. Horopian; M.M. Foulds (1979). Neurotoxic fragrance produces ceroid and Myelin Disease. *Science* **20**, 633–635

16. Plassche, E.J. van de; F. Balk (1997). Environmental risk assessment of the polycyclic musks AHTN and HHCB according to the EU-TGD. National Institute of Public Health and the Environment (RIVM), Report 601503008, Bilthoven

17. Schwartz, S.; V. Berding; M. Matthies (1999). Aquatic Fate Assessment of the Polycyclic Musk Fragrance HHCB – Scenario and Variability Analysis in Accordance with the EU Risk Assessment Guidelines. *Chemosphere* (accepted)

18. Schwartz, S.; V. Berding; M. Matthies (1999). Umweltexpositionsabschätzung des polycyclischen Moschus-Duftstoffes HHCB – Szenarienanalyse mit EUSES. *Umweltmed. Forsch. Prax.* **4**, 7–11

19. AWWR (1997). Ruhrwassergütebericht. The study group of the waterworks at the Ruhr (AWWR) and Ruhrverband, Essen

20. Berding, V.; S. Schwartz; M. Matthies (1999). Visualisation of the Complexity of EUSES.*Environ. Sci. & Pollut. Res.* **6**, 37–43

21. Berding, V.; S. Schwartz; M. Matthies (1999). Scenario analysis of a Level III multimedia model using generic and regional data. *Environ. Sci. & Pollut. Res.* (submitted)

22. Feijtel, T.C.J.; G. Boeije; M. Matthies; A. Young; G. Morris; C. Gandolfi; B. Hansen; K. Fox; M. Holt: V. Koch; R. Schröder; G. Cassani; D. Schowanek; J. Rosenblom; H. Niessen (1997). Development of a Geography-referenced Regional Exposure Assessment Tool for European Rivers – GREAT-ER; *Chemosphere* **34** (11), 2351–2374

23. Struijs, J. (1996): SimpleTreat 3.0: a model to predict the distribution and elimination of chemicals by sewage treatment plants. Report No. 719101025. National Institute of Public Health and Environmental Protection (RIVM), Bilthoven, The Netherlands

24. Boeije, G. (1999). Chemical fate prediction for use in the geo-referenced environmental exposure assessment, PhD thesis, University of Gent, Belgium

25. Dyck, S. (1980). Angewandte Hydrologie, Teil 1 – Berechnung und Regelung des Durchflusses der Flüsse. Ernst, Berlin, München, Germany

26. Bundesanstalt für Gewässerkunde (1993). Jahrbuch Bundesrepublik Deutschland, (German Hydrologic Yearbook, Federal Institute of Hydrology) Koblenz, Germany

27. Round, C.E., A.R. Young; K. Fox (1998). A regionally applicable model for estimating flow velocity at ungauged river sites in the UK. *Journal of the Chartered Institution of Water and Environmental Management* **12** (6), 402–405

28. Ruhrverband (1999). Basic information on sewage treatment plants of the Ruhrverband – Personal communication. Ruhrverband (Institution for River Ruhr Water Management), Essen

29. NRW (1994). Temes Jahresberichte, Tabellenband 1994. Landesumweltamt Nordrhein-Westfalen, Düsseldorf

THE FUTURE OF THE ENVIRONMENTAL MODELING IN RISK ASSESSMENT IN SLOVENIA

Country Report

BORIS KOLAR
Laboratory for ecotoxicology and biotechnology
Environmental protection institute,
Public health institute Maribor,
Prvomajska 1, 2000
Maribor, Slovenia

Abstract

Recently a law was passed in Slovenia on chemicals that gives a legal foundation for evaluating chemical risk. Following special training a group of experts from the field of medicine and environment protection began to assess the risk of selected groups of chemicals according to the EUSES programme. In the future the risk assessment for chemicals will be harmonised with the EU legislation, and for this reason we will build up a suitable legislation in Slovenia.

We will form an information network, which will include data on production, import, transport and use of chemicals. The aim of training experts is equal rights for co-operation in the preparation of risk assessments for chemicals in the EU. On a regional level we will form model risk assessments which will supplement the EUSES programme and they will take into consideration the particularities of the Slovene area.

1. Introduction

Slovenia is a small central European country with an area of 20.256 km2 in which live two million inhabitants with a gross local product of US $ 9161 in the year 1998. It has a predominantly mountainous and hilly relief, with an average above sea level height of 500m. A special thing about Slovenia is that here Central European and Mediterranean natural characteristics meet and interweave. Slovenia is open on all sides: to Central Europe, towards the West, to the Balkans and towards Eastern Europe. This means an outstanding landscape and one of the most diverse biotic regions in Europe. In the flat land of Slovenia, which covers only a fifth of the country, 80% of all economic activities take place. Here the environmental problems have accumulated because of

J.B.H.J. Linders (ed.), Modelling of Environmental Chemical Exposure and Risk, 223–225.

industrialisation, urbanisation and intensive agriculture especially in the second half of this century. Demographically the hinterlands are emptying, agricultural land is being overgrown by forests that today cover more than 54% of the country. Almost half of Slovenia lies on the Karst region.

2. Subterranean water – mirror of the environment's activities in the past

Slovenia is rich with subterranean, in the Karst region underground, waters that represent a quality source of drinking water. In the flat land areas, which in Slovenia lie predominantly on alluvial deposits, agricultural use of pesticides and nitrates are a threat to the shallow subterranean waters. Also a number of illegal dumping grounds for municipal and industrial waste as well as wastes from the production and use of pesticides represent another source of subterranean water pollution. A result of increased economical activity is also increased sea, road and railway traffic and the environmental risks linked to them. Individual segment conditions of the environment are monitored either on a national or local level. On the national level there are: monitoring of air quality, which was already started in the sixties, monitoring of surface and ground water which has been done since the second half of the eighties. In 1998 constant control on the content of pesticides in food began.

3. Risk assessment for chemicals in Slovenia today and how to carry on

Slovenia began using the methods of risk assessment for chemicals as tools for assessing their effect on humans and the environment relatively late. It was only when the law on chemicals, which came into effect on 29.05.1999, accounted for the legal foundation for risk assessment of chemicals. In 1999 a group of experts was set up, who are preparing the risk assessment for carbon tetrachloride, atrazine, chlorpicrin, methylbromide, paraquat, 2,4,5-T. The assessment is being done with the help of the EUSES programme. Slovenia is also preparing to join the pesticide aquatic risk indicators project.

In the process of approaching European integration it is in Slovenia's interest to train expert cadre who, on equal footing and with complete responsibility, will be capable of taking on a part of the task in the field of risk assessment of chemicals in the EU.

In regional (national) dimension the aims and future policy in the field of risk assessment for existing and new chemicals are the following:

- **To pass missing legal documents**; particularly acts, which regulate the handling of chemicals in Slovenia.
- **To form an effective information network;** experience from some other West and Central European countries that are in the phase of joining the EU show that there are often extensive data bases available in the field of environment, nutrition and human health that can be effectively included in risk assessment. In Slovenia

we will set up an information network through which it will be possible to obtain data on existing data bases - on performing monitoring programs in individual segments of the environment, accidents with chemicals in the environment, poisoning of people and similar. It will be possible to evaluate and use the data for risk assessment. A special part of the information network will be the database on import, production and traffic with chemicals.

- **To select risk assessment models and prepare scenarios;** with which it will be possible to fill gaps that occur in the use of the EUSES programme, especially the limited flexibility of the programme. In risk assessment we wish to take into consideration all the particularities of Slovenia, and to preserve them in the future.
- **To continue with training of expert cadre;** the risk assessment for selected chemicals runs according to the EUSES programme. The aims of training are above all based on this programme.

4. Conclusion

Slovenia was relatively late in adopting the methods and tools accepted by the EU for risk assessing of chemicals. Since passing suitable legislation activities are running at an accelerated rate. With these activities Slovenia wishes to join countries that can assure relevant risk assessment at home as well as for the needs of the EU.

ENVIRONMENTAL EXPOSURE OF PLANT PROTECTION PRODUCTS

Portuguese experience

FLÁVIA ALFARROBA
Direcção-Geral de Protecção das Culturas
Quinta do Marquês
2780-155 OEIRAS
Portugal

Abstract

Environmental exposure models play an important role in the risk assessment of plant protection products. The adoption of the EU legislation concerning the placing of plant protection products on the market - Directive 91/414/EEC - improved the use of models in the regulatory decision making process [1]. To harmonise and to give orientation on the use of exposure models within European Union Member States a very important contribution has been given by the FOCUS groups.
Portugal has followed all these developments and the Portuguese experience is presented.

1. Introduction

The assessment of environmental risks of plant protection products (PPP) has to consider both the effects as well as the extent of exposure. Quantitative exposure estimates are, therefore, needed to perform acceptable risk assessments. Because of the increasing importance of risk assessment in the registration process of PPP, the requirements for environmental exposure have been and are still increasing.
The Council Directive 91/414/EEC of 15 July 1991 concerning the placing of plant protection products on the market, that came into force in July 1993, laid the foundation for a harmonised system of registration of those products within the EU. The Directive, through the different Annexes, which followed, has given a great importance to the use of mathematical models. Calculation results for Predicted Environmental Concentrations (PEC) are taken into consideration either as criteria for a decision on the need for conducting further experiments or as a support for the evaluation and decision making process. Therefore, PEC calculations need to be performed for soil to assess effects on macro and microorganisms; for surface water to assess effects on aquatic organisms; for ground water, to evaluate the potential effect on drinking water; and for air [2].

J.B.H.J. Linders (ed.), Modelling of Environmental Chemical Exposure and Risk, 227–230.

Since the regulatory use of exposure models was new, there were neither clear or detailed guidelines nor a generally agreed practice on how they could be used and how the results could be interpreted.
In order to solve those problems within the EU countries and to help the users/regulators in using models a "FOrum for the Co-ordination of pesticide fate models and their USe" (FOCUS) was established in 1992. From this time onwards an excellent and very useful work was done by the different sub-groups involved (ground water, surface water and soil) in assessing and selecting the most appropriate models for each compartment, discussing the respective model limitations, deficiencies and problems related with their validation and indicating the list of available models that can be used to address EU registration requirements. More recently, a set of standard European modelling scenarios is being developed to account for the climatic and environmental differences among the EU countries [3].

2. Experience in Portugal

In Portugal, the registration of PPP started in 1967 when the respective law came into force. From this time until the adoption of the EU legislation on PPP the evaluation of those products followed the data requirements internationally agreed and the decision making process was based on hazard assessment rather than on risk assessment. Thus, only effects were taken into consideration and not exposure, excluding a few cases of direct application of PPP in water as it was in the case of rice cultures or in irrigation canals in which calculations based on equations using several simplified assumptions were used.
The international developments in the risk assessment and the Toxicity Exposure Ratio (TER) approach - ecotoxicological effects/predicted environmental concentration - followed in the EU decision-making process gave the opportunity to go deeper into the exposure and risk assessment. Portugal has been involved in the discussions of these subjects at EU level since the beginning and has followed the FOCUS activities through the participation in the FOCUS Steering Committee.
In the Portuguese risk assessment a tiered approach is followed and exposure estimates - PEC values (initial and time weighted average concentrations) - are calculated, according to the recommendations of FOCUS sub-groups [4], [5], [6], using standard scenarios and simple models based on theoretical equations and tables (spray drift tables, crop interception tables), basic laboratory properties (e.g. DT_{50} and K_{OC} or K_{OM} values) and the conditions of application (dose/rate, number of applications and time between applications).
Since the results obtained are mostly based on worst case assumptions they are considered a good decision-making tool for a generic low tier worst case assessment and in many situations there is no need to go further in the risk assessment.
However, if the results give rise to concern (indication of leaching potential, high risk to aquatic or terrestrial organisms) other more specific approaches are needed. In this case, the results of the low tier assessment are used as a trigger for semi-field or field studies which have to be carried out in similar and comparable agronomic, climatic and environmental conditions of Portugal. As an example, exposure estimates from leaching

models using standard scenarios (PELMO, PESTLA models) are used as a trigger for field or lysimeter studies and in a higher tier approach, if considered necessary, for monitoring studies.
From the ecotoxicological point of view, when PEC-s give rise to unacceptable margins of safety the refinement of the risk assessment requires that higher tier studies, observing the semi-field or field effects on aquatic or terrestrial organisms under realistic conditions of use, be performed.
Nevertheless, if higher tier levels of exposure assessment have to be considered, complex mathematical models using specific scenarios and site specific data for South European countries have to be applied and the necessary data needs to be available. Until now, no environmental exposure models that could be used in the registration process were developed in Portugal and, in general, little work was produced in this field in the other South European countries that could account for the peculiarities of the Mediterranean area. Efforts should be made to fill these gaps. As an example, the EU South European countries started working on a common approach and criteria for a proper environmental fate modelling, exposure scenarios and risk assessment concerning the use of PPP on rice, a typical Mediterranean culture.

3. Conclusions

Environmental exposure models are a useful tool to improve the quality of risk assessments. In the risk assessment of plant protection products performed by the Portuguese authorities, a tiered approach is followed, starting with simple calculations and standard scenarios, some of them developed to take into account the climatic and environmental conditions of Central and North part of Europe. For a generic worst-case estimation this is considered an acceptable approach.
When it is necessary to go further in the risk assessment low tier exposure estimates are used as a trigger for semi-field or field studies to be performed under realistic conditions of use of PPP in Portugal.
However, particularly for higher tier levels, the risk refinement should become closer to more realistic agronomic, climatic and environmental conditions. Thus, specific scenarios and site-specific data from South European countries need to be used and should be developed.

4. References

1. EEC (1991) - Council Directive 91/414/EEC of 15 July 1991 concerning the placing of plant protection products on the market. O.J. N° L230, 19.08.1991, 31 pages.

2. FOCUS (1995) - Modelling Environmental Fate of Plant Protection Products in the Context of their Authorisation within the European Union. European Commission Document 1694/VI/95. Brussels, 6 pages.

3. Linders, J. (1999) - Report FOCUS - Workshop on Soil/Leaching and Surface Water Scenarios. 3 and 4 September 1998, RIVM, Bilthoven, NL.

4. Adriaanse, P.; Allen, R.; Gouy, V.; Hollis, J.; Hosang, J.; Jarvis, N.; Jarvis, T.; Klein, M.; Layton, R.; Linders, J.; Schäfer, H.; Smeets, L. & Yon, D. (1996) - Surface Water Models and EU Registration of Plant Protection Products. European Commission Document 6476/VI/96, Brussels, 218 pages.

5. Boesten, J.; Businelli, M.; Delmas, A.; Edwards, V.; Helweg, A.; Jones, R.; Klein, M.; Kloskowski, R.; Layton, R.; Marcher, S.; Schäfer, H.; Smeets, l.; Styzcen, M.; Russell, M.; Travis, K.; Walker, A. & Yon, D.; (1995) - Leaching Models and EU Registration. European Commission Document 4952/VI/95, Brussels, 123 pages.

6. Boesten, J.; Helweg, A.; Businelli, M.; Bergstrom, L.; Schäfer, H.; Delmas, A.; Kloskowski, R.; Walker, A.; Travis, K.; Smeets, L.; Jones, R.; Vanderbroeck, V.; Van der linden, A.; Broerse, S.; Klein, M.; Layton, R.; Jacobsen, O-S & Yon, D.; (1996) - Soil Persistence Models and EU Registration. European Commission Document 7617/VI/96, Brussels, 74 pages.

ENVIRONMENTAL CLEANUP PROGRAM IN HUNGARY

Hungarian Report

EVA DESEO
Ministry for Environment,
H-1394, Budapest
P.O.Box 351
Hungary

Abstract

By the end of the twentieth century, most countries recognised that the environment had undergone significant environmental damage, and they made the first steps to identify and eliminate various causes of pollution. The first experience in remediating environmental damage was directed at former Soviet military bases, and was begun in 1991 after Soviet troops left Hungary. Countrywide co-ordination was necessary in setting up the list of priorities and identifying, assessing, and characterising polluted areas. This national program was begun within the Ministry for Environment. To co-ordinate the program, the Environmental Remediation Program Office was created in the Institute for Environmental Management.

1. Introduction

1.1. INTERNATIONAL EXPERIENCE

By the end of the twentieth century most countries recognised that the environment had been seriously damaged. About the seventies the first steps were made to identify and eliminate various causes of pollution. In the United States the remediation experience has involved at the Superfund sites, where an individual risk analysis is prepared for each examined area. The specific tasks to be carried out depend on the result of these risk analyses [13].

It was only 10-15 years ago that the developed West-European countries realised that polluted areas had to be gradually cleaned up. The resolution of this problem needs a long-term well-established program. Germany, the Netherlands, Denmark were among the first to react to the problem, while a high level program is currently being implemented, or planned in almost all of the EU member states. The program

J.B.H.J. Linders (ed.), Modelling of Environmental Chemical Exposure and Risk, 231–240.

organisers have faced numerous issues that have given rise to serious social debate: determining responsibility, financing the work, developing the criteria for establishing intervention priorities and the rational objectives of interventions. Most countries today tend towards using limits in the first phase of uncovering pollution and individual risk analysis in detailed investigations.

There are many methods of financing and government budgets are dominant, but the differences lie in the kinds of income schemes that provide coverage. These include product charges, waste taxes, and environmental protection contributions and fines.

1.2. HUNGARIAN BACKGROUND

Hungary is a small country in Middle Europe, in the Carphatian Basin. Its area is 93.000 km^2, there are about 10 million inhabitants, and density of population is near to 100 persons per km^2.

The Hungarian regulation for the use and protection of environmental elements has to be adjusted gradually to the EU's demands. It is not only the Hungarian situation, but also the aim of all the associated states. This demand will even be existing after the state will become member-state of the EU [1].

Economic growth, especially vigorous industrial development took place in Hungary without the constraints of strong environmental protection regulation up to the end of the seventies and beginning of the eighties. Although the legal regulations concerning environmental protection later caught up with contemporary requirements, compliance with the regulations fell far short of the theoretical strictness of limits. By the reason of the economical difficulties of this time only insubstantial amounts could be spent on environmental protection, and this situation led to a gradual accumulation of non-degradable and slowly degrading pollutants in groundwater and the soil.

In Hungary the environmental problem has been recognised about ten years ago. Hard preparation work precedes the start of this program.

By estimation there are approximately ten thousand polluted areas in Hungary, where remediation would have been an imperative for years, even decades. The first experience in Hungary in remediating environmental damage was directed at former Soviet military bases. The government's 1991 short- and medium term action plan, which identified the tasks of surveying, assessing, and eliminating accumulated environmental pollution - presented by abandoned Soviet barracks and training grounds after they had left Hungary - can be considered as the starting point for the Remediation Program. After the investigation of the 150 sites abandoned, half of them needed immediate cleanup intervention. The remediation of the most polluted of the former soviet properties will have been completed in one to two years.

The experiences obtained in the course of privatisation, the revival of the real estate market and the development of public participation in environmental protection helped provide justification for the Ministry for Environment and Regional Policy's original initiative.

The government launched the National Environmental Remediation Program in 1996 in order to asses polluted areas, uncover damage that falls within the scope of the government's responsibility, and eliminate the damage.
In September of 1997 Parliament has approved the National Environmental Program which contains the Remediation Program.

2. Strategy of the program

The main environmental elements to be investigated in this program are soils and subsurface water. In Hungary, promoting and safeguarding the quality of groundwater is extremely important because groundwater provides 90% of Hungary's drinking water needs [3].
Countrywide co-ordination was necessary in setting up the list of priorities and identifying, assessing, and characterising polluted areas. After two years of preparation this national program was begun within the Ministry for Environment.
The budget of this program is annually about 5 million USD. To co-ordinate the program, the Environmental Remediation Program Office was created in the Institute for Environmental Management. This program deals with the nation-wide inventory of pollution sources and polluted sites. Its aim is to create a homogeneous database of sites for remediation, to calculate the national priority list of polluted sites, but for the remediation itself of these polluted sites a different authority will be responsible. This program deals only with the special cases, when the responsibility for the pollution could not be established. In Hungary programs dealing with environmental remediation arrive to a budget of about 30 million USD yearly.

3. Government responsibility

The new environmental law stipulates that, if no other person can be made responsible, it is the task of the government to eliminate the consequences of significant environmental damage. The pollution must be considered as government's responsibility, when the polluter is unknown, (or the presumed polluter cannot be proved to be responsible). There are cases when the firm has been terminated without a legal successor, or the liquidation is going on and the liquidation assets have been proved to be insufficient for cleaning up the environmental damage. In such cases the given area must be cleaned up in the framework of the Remediation [4].
It is also the responsibility of the government to clean up long-term environmental damage caused by governmental budget agencies. In the framework of the Remediation Program so-called subprograms are developed with the co-ordination of the Program Office. These subprograms have two directions. At first data available are collected into the nation-wide uniform database - after inventory, and national priority calculations - the responsible agency has to organise its remediation subprogram, and the financing of

it. The developing of this subprogram - based on data collection of polluted sites - has been financed by the Remediation Program.

4. Connection to other programs

4.1 NATIONAL ENVIRONMENTAL PROGRAM

The National Environmental Program - which is approved by the Parliament – serves for uniform structure for these different actions.

4.2 WELLFIELD PROTECTION PROGRAM

The remediation Program has connection to other environmental programs too. The Wellfield Protection Program adopted by the Government in 1995 is aimed at securing both operating and prospective wellfields located in vulnerable environment. In the protection zones of wellfields - determined in this program - potential pollution sources are listed, and monitoring zones are created. Decisions concerning the measures that must be taken in order to secure an area are made in consideration of the cost/benefit analysis of each alternative. This can also include elimination of the existing pollution in the soil or in the groundwater as well as eliminating the cause of the pollution. The need for the co-ordination between the two programs has been recognised in an early stage, the exchange of information is current and decision making is harmonised.

4.3 NATIONAL ENVIRONMENTAL HEALTH ACTION PLAN (NEHAP)

The National Environmental Health Action Plan has also established a database for the polluted areas for evaluating the environmental hygiene risk [11] and considering local characteristics and possibilities. In some study areas detailed results provide a reasonable good basis for comprehensive priority calculation, which is useful for the Remediation Program too.

4.4 CLEANUP PROGRAM OF THE HUNGARIAN RAILWAY COMPANY (MÁV RT.)

This program has started in 1997 in good relation with the Cleanup Program. All of the registered polluted sites in the property of this company (railway stations, workrooms, embankments) are incorporated in the database of the Remediation Program, priorities will be calculated and will be considered in the order of succession in the cleaning up processes.

4.5 MINING STRUCTURE CONVERSION PROGRAM (SZÉSZEK)

This program has started earlier, based on the Government's decision of 1991. It has also a lot of relations with the remediation program. The database will be incorporated in the database of polluted sites as well. About 1000 sites are registered, but the most considerable ones are remedied already.

4.6 INDUSTRIAL PARK PROGRAM

The Industrial Park Program has only begun in 1997. Former industrial zones will be reconstructed. In the Mining Structure Conversion Program new industrial parks have been created by the rehabilitation of old mining properties. It will be very important to incorporate the maximum available data of the old, polluted industrial sites.

4.7 MILITARY CLEANUP PROGRAM

This program is based on the preparation made in the framework of the remediation Program. The program is financed since 1998 by the budget of the Ministry of Defence.

5. Tasks

5.1 GENERAL TASKS

Research, developing the legal, technical and information technology system has been determined a major general task, which includes the compilation annually of the priority list.

The Ministry for Environment has recently prepared a legislation draft, which contains the provision of the groundwater directive (80/68/EEC). It deals – among others – indirect discharge of pollutants into the groundwater, and in this respect it gives some provision about the soil protection too [1]. It is a real demand in Hungary to regulate the environmental criteria of good chemical status of groundwater, as it is one of the goals of the draft water framework directive (COM/97/49).

Important general tasks are management, co-ordination, and operation of the program, which is organised by the Remediation Program Office.

5.2 NATIONAL TASK

The main national task is the nation-wide pollution source and polluted site registration. The development of subprograms for remediation projects, for which government

organisations bear statutory or contractual obligations, is an important part of this registration.
Registration of long-term pollution into the property registers [2] is another important task, which includes the comprehensive nation-wide survey of pollution sources, the development and operation of a national monitoring system, the follow up, the efficiency analysis, and the implementation of subprograms, the publication of information about the program, and finally the education.

5.3 SPECIFIC TASKS

As a specific task can be mentioned the announcement of tenders for companies relating to direct environmental remediation operation on individual sites, including site assessment and feasibility study, emergency measures, environmental remediation (concentration, collection, and removal of pollutants), and follow up studies.

6. Program phases

6.1 SHORT-TERM PHASE

The first two years (1996 and 1997) can be considered the program's short-term phase. The development of the research, information technology, regulatory, and monitoring system started. During this period the program was established and its methodology created. The government's responsibility and participation were clarified. The program's medium-term phase was compiled. The process of nation-wide registration has begun. Individual tasks were carried out, as well as emergency measures, investigations and site remediations.

6.2 MEDIUM TERM PROGRAM

Tasks relating to the whole program continue from 1998 to 2002. The information technology background, research and technological development, designing and implanting the financing scheme of the program should be carried out. These include compiling and publishing a list of the most suitable modern technologies for cleanup projects as well as developing methodologies for risk evaluation and cost/benefit calculations. Developing a comprehensive assessment – by a registration system so-called KÁRINFO – which will be integrated in the Environmental Information System, will result in the most comprehensive version of the Remediation Priority List for 2002.

6.3 LONG-TERM PROGRAM

Final long-term phase beginning in 2003 will ensure a continuous maintenance of data collected. Individual tasks will be decided yearly and carried out by the co-ordination of Program Office.

7. Results

7.1 RESULTS IN THE SHORT TERM PROGRAM (1996-1997)

7.1.1 Research, developing information system

The proposed Hungarian legislation is to set up a system of the environmental criteria and limit values of groundwater and soil as the followings: background values (A), threshold values of pollution (B), threshold values of measure (C), target values (D). The complete system of the limit values “A”, “B” and “C” [5] were established by the hard work of an expert group, which took into consideration the values, which can be found in the Dutch and German lists, as well as the Canadian values and guidelines issued by the US EPA. The threshold values and target values are different depending on the vulnerability of the aquifers [1].
In the short-term phase of the program the methodology was created. Soil protection methodologies are determined [7, 8], use of transport models for groundwater contamination modelling [12] are published in a handbook.
Questionnaires and software are developed for calculating the priority number of the registered polluted sites for decision making. The priority number is based on preliminary risk assessment.
The responsibility and participation of the government [4] were clarified. The program’s medium-term phase was compiled. Monitoring system’s conceptions [9, 10] were determined.

7.1.2 Nation-wide inventory process

The process of nation-wide inventory begun. As the first step of the process the regional environmental authorities fulfilled the questionnaires distributed by the Remediation Program Office. This preliminary database contains 173 polluted sites. Fig. 1 shows the distribution of the main pollutant material of polluted sites registered in the short-term phase.

7.1.3 National Priority List of Cleanup

Based on the collected data of the above mentioned 173 polluted sites – using the preliminary priority (PRIOLIST) program – sites could be chosen for most urgent intervention.

7.1.4 Individual operations

Operations on polluted sites in the short term: The Program Office is in the process of preparing bids to solicit remediation activities [6]. It is envisioned that the various phases of the cleanup operations may include:

- I. Site assessment and feasibility study: in 1996 fifteen sites, and in 1997 nine more sites (those with the highest priority) were selected for investigation. Seven sites showed no considerable environmental damage (having done on these sites emergency measures). The study of site assessment contains the detailed environmental risk assessment.
- II/A. Emergency measures: 13 sites (in 1996 eight, and in 1997 the other five) required emergency measures. After taking the most urgent steps to prevent additional environmental damage, site assessment and feasibility studies were performed.
- II. Rehabilitation of the environment (soil and groundwater) until specified contamination limits are reached. Based on the result of the site assessment of the 24 investigated sites, seven of them did have no considerable environmental damages. On six polluted areas cleanups have begun, two of these already have been finished.
- III. Monitoring system development and operation of efficiency control of the cleanup process. On six sites of the 24 sites a monitoring system has to be installed, for all together 24 shallow groundwater wells. Regional Environmental Authorities will operate them.

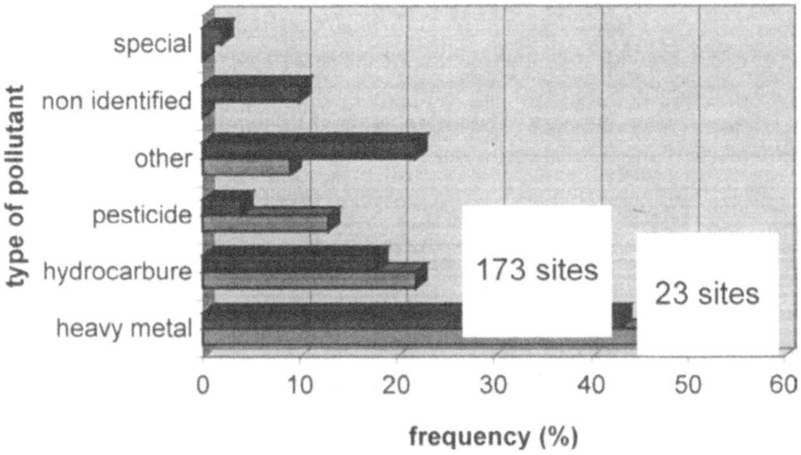

Figure 1. Distribution of different types of pollutant

7.2 RESULTS IN THE MEDIUM TERM PROGRAM (OF 1998-2002) IN THE FIRST YEAR

7.2.1 Research, developing information system

The main aim of this phase is to identify, assess and characterise all of the pollution sources and polluted sites with a countrywide co-ordination, and to set up the priority list of polluted areas. Parallel to the practical experience there is a permanent need to renew and verify the conception.

7.2.2 Nation-wide inventory process

About 25,000 potential pollution sources and polluted sites have been identified and set into the database. These data have been collected by subprograms from different studies. These data are not sufficient for priority calculations. Some data concerning the environment and the pollution give opportunity to make a screening, for choosing the important polluted sites. For these sites questionnaires have to be fulfilled.

7.2.3 National Priority List of Cleanup

Using the actual complete database a National Priority List of Cleanup will be calculated annually.

7.2.4 Individual operations

The cleanup of sites, which had begun in the short-term phase, is in progress. Most cleanups will be finished by the end of 1998 or 2000 while one of them will be postponed to 2002.

8. **Conclusion**

This program is an important result of the last decade concerning the improvement of the environmental condition. The pollution of the soil and the groundwater is obviously less perceptible than smoke-emitting chimneys of factories, petrochemical city smog, or dead fish in an oil-polluted river. But the human health can be threatened by the consumption of polluted water or garden-grown vegetables or even by recreation in polluted areas. Pollutants do get washed off into surface waters, and decomposing hazardous wastes threat the environment through the air. This long-term environmental damage constitutes one of the factors of environmental pollution that has an unfavourable effect on public health and, ultimately, life expectancy.

The experience obtained by individual remediation studies of this program gives an excellent opportunity for finding an experimental base for controlling of theories. The application of different transport models, experimental methodology of the degradation of polluting substances in different environmental compartments.
Adaptation of the most suitable technology may reach a considerable cost efficacy. In this way from the same budget much more sites could be remedied.
This program will be finished only in the next century, but hopefully the experience gained in the past, will render the program more and more efficient.

9. References

1. Balashazy, L., 1998. Regulation tasks for the protection of*groundwater Hungarian Hydrologic Society Conference Proceeding Jul. 5-8.*

2. Burger, F., 1997. Registration of long-term environmental damage into the property registers. *Guide for Remediation Vol.1.* (ed. by Hungexpo)

3. Deseo, E., Deak, J., 1997. Groundwater quality in Hungary - a general view - in freshwater Contaminant Proceeding of Rabat Symposium S4. IAHS Publ. No.243. (227-234)

4. Docs, B.Zs., 1997. Responsibility for environment, remediation governmental of environmental damages. *Brochure of Remediation Vol. 1.* (ed. by Hungexpo)

5. Eötvös, K. 1998. Soil and groundwater remediation limit values.*Guide for Remediation Vol.2.* (ed. by Hungexpo)

6. Galgoczi, I., 1997. Short presentation of projects started in the short-term phase (1996-1997) of the Remediation Program. *Brochure of Remediation Vol. 3.* (ed. by Hungexpo)

7. Horvath, Zs., 1997. Soil protection. *Brochure of Military Environmental Protection Vol. 3.* (ed. by ZMNE Military Environmental Security Centrum Budapest)

8. Kadar, I., 1998. Soil remediation technologies models *Handbook of Remediation Vol.2.* (ed. by Hungexpo)

9. Kelly, V.E., and Bogardi, I., 1991.Designe, operation, and some examples of field monitoring networks. In: *Hydrogeological Basis of Ecologically Sound Management of Soil and Groundwater.* (ed. by H. P. Nachtnebel & K. Kovar)(IAHS Publication **202**), 257-268.

10. Liebe, P., 1998. Groundwater monitoring on long-term polluted areas. *Guide for Remediation Vol.2.* (ed. by Hungexpo)

11. Pinter, A., Horvath, A., Szabo, Z., Dura, G., Horvath, V., 1998. *Current practice of health risk assessment of soil pollution in Hungary Central European Journal of Public Health Vol. 2/6.1998. (144-147).*

12. Simonffy, Z., 1998. Application of pollution transport models*Handbook of Remediation Vol.1.* (ed. by Hungexpo).

13. Szvetnik, N., 1997. International experience in remediation of polluted sites. *Brochure of Remediation Vol. 2.* (ed. by Hungexpo).

MODELLING EXPERIENCES IN THE SLOVAK REPUBLIC

Country Report

MARTIN MURÍN
Ekotoxikologické centrum Bratislava s.r.o.
Nádražná 36
900 28 Ivanka pri Dunaji
Slovak Republic

In the Slovak Republic limited experience is available in the field modelling of environmental chemical exposure and risks. In separate areas some of modelling aspects are used e.g. the estimation of concentrations of specific chemicals from waste water discharges, ambient air pollution from point sources, specific studies.

Technical guidelines for exposure and risk assessment are under preparation in connection with new chemicals legislation cf. European Guidelines for new and existing chemicals.

Limited experiences are also available in connection with bilateral or multilateral co-operation. Different projects in this field were done in past time from which we can mention the following examples:

- **Risk assessment / Risk Management Demonstration Project in the Slovak Republic, Žilina Case Study. (1995 - 1997)**
  Co-operation between US EPA and Slovak ministry of Environment (1994-1997), which have resulted in some reports:
  "Occurrence of Nickel in Environmental Compartments of Žilina Town and Surrounding Area"
  "Assessment of Risks of Industrial Chemicals from Point Sources of Eastern Industrial Zone in Žilina Town"

- **Ecotoxicological Methods and Strategies for Environmental Impact Assessment (1992-1997)**
  Co-operation between the Danish and Slovak ministries of Environment, which led to the publication of the report:
  "Priority Setting of High Production Chemicals in the Slovak republic"
  and a publication in the scientific literature:

J.B.H.J. Linders (ed.), Modelling of Environmental Chemical Exposure and Risk, 241–242.

"Aquatic Hazard and Risk Assessment of Two Selected Substances Produced in High Volumes in the Slovak Republic"; Chemosphere, Vol. 34, No.1, pp. 179-190,1997

- **Ecological Risk Assessment of Pollution by (heavy) Metals and Organic Micropollutants in the Danube Catchment Area**
  Multilateral project, Phare, 1995 - 1998
  Publication has been taken place in 4 phases reports.

It is difficult to evaluate modelled and measured data on the national scale. The limited experience shows positive tendencies.

The systematic evaluation / monitoring of exposure concentrations are mostly focused on heavy metals and other „traditional" pollutants like DDT, lindane and others.

Initial work has been done in the area of high production volume chemicals (HPVC) relevant for the Slovak Republic based on an inventory done in the years 1992-1993. It was recognised that risk assessment including modelling of exposures should be incorporated into the decision-making process.
This could be presented by new guidelines of the Slovak Ministry of Environment - Risk assessment methodology; Risk assessment for river bottom sediments.
Both guidelines partly discuss also the use of models in different steps of hazard and exposure assessment.

Different modelling tools are used in specific areas like approximation of outlet concentrations in wastewater treatment plants, distribution of pollutants from point sources, estimation of spreading pollution in ground water and others.

New questions are set in the approximation process of the Slovak legislation with the European Union. Broader co-operation between different sectors in the country and internationally are the only possible ways to increase the efficiency and the advantages of models in the environmental protection.

DOJRANSKO EZERO (DOJRAN LAKE) PROGRAM 1999-2001

VLADIMIR KENDROVSKI
Republic Institute for Health Protection-Skopje,
Skopje
Republic of Macedonia

Abstract

This Program has three major elements:

1. Identifying the primary threats to human and ecosystem health within the watershed.
2. Involving the people most likely to be concerned or most able to take action to solve the problems.
3. Taking corrective actions in a comprehensive, integrated manner once solutions are determined.

The Republic Institute for Health Protection (RIHP) is working across the Republic of Macedonia to establish watershed protection using procedures developed and used by EU and EPA-USA.
The results will be used for the risk assessment of water contamination and impact to the human health.

1. Background

The Dojran Lake has a higher trophic level than other similar lakes in Europe. It has a watershed area of 270 km^2, with a maximum depth of 11 m. The entire lake at the Macedonian side is surrounded by 4 places: Star Dojran, Nov Dojran, Sretenovo and Nikolik with 2325 inhabitants; there are also tourist areas: Mrdaja, Achicot with 40 hotels and accommodation places.

2. Problems

Water quality problems in Dojran Lake first became evident after 1988/89 when it has been used for irrigation and agricultural uses and after a dry period with a decrease of water quantity. Between 1988 -1998 the volume reduction was most noticeable with -200 cm under the 0 level. There appears massive toxic cyanobacterial blooming every

J.B.H.J. Linders (ed.), Modelling of Environmental Chemical Exposure and Risk, 243–244.

year in the water in the Dojran Lake and aquatic life and recreation are suffering from excessive inputs of it. Cyanobacteria present water-borne hazards to health via drinking water and recreational water.

3. Participants

Studies and investigations will be provided in the field and in the laboratory. The scope of the work will be: preparation of sampling sites, screening of the site, monitoring, assessment of the impacts of the cyanobacterial toxins in the environment with a general schedule of activities and with participant of RIHP, IHP-Veles LD Gevgelia, Veterinary Institute, Republic Hydro-Biology Institute. In the Project, we will study the rate of cyanobacterial toxins and their ecotoxicological properties. With standardised tests we will investigate the ecotoxicological properties of blooms trough the seasons during a 3-year period.

4. Goals

The main goal is to give an impulse to the development of Public Health System, which actively responds to the needs of the community. The aim of the specific study is to monitor the toxin mobility in surface water and sediments.

5. References

1. Bell SG, Codd GA. (1995). Cyanobacterial toxins and human health. *Rev Med Microbiol*; 5: 256-64.

2. WHO (1997). Report of the Working Meeting on Chemical Substances in Drinking Water, Geneva, 52 (abstr).

SOME METHODOLOGICAL ASPECTS OF SOIL DATA RECEIVING AND USE FOR THE ENVIRONMENTAL PROGNOSIS

Country Report

GALINA V. MOTUZOVA
Moscow State University,
Faculty of Soil Science.
119899, Vorobiovy Gory,
Moscow, Russia.

Abstract

The important position, which soil takes in the ecosystem and the importance of soil processes for the transformation of the substances in the environment provide the high demands to validation of soil parameters, which are used in modelling the environmental risk of chemicals. The advantages and disadvantages of some methods (empirical analogies, technological, geophysical, field experiments and laboratory physico-chemical imitation) of the measurement of soil parameters are examined. The examples of use of field and laboratory experiments for establishing soil parameters are discussed from the point of view of their validation and effectiveness for environmental prognosis.

1. Prognosis of Environmental Risk

At the present time the concept of environmental risk is dominant in forecasting the environmental concentration of chemicals and it has excluded or changed many others approaches. The main idea of this concept is rather clear expounded; this circumstance has provided the opportunity of mathematical modelling of environmental chemical exposure and risk.

The degree of adequacy of models depends on their methodological basis. The methodological problems are the most important in the mathematical modelling of any events, especially of those, which are directed on the environmental prognosis. All models, which describe the behaviour and distribution of different kinds of substances (pesticides, heavy metals, acid rain, oil and oil products and so on) in the environment and the environmental consequences of pollution of the ecosystem obligatory include «soil block» because soils play the key role in the way substances are distributed in the

J.B.H.J. Linders (ed.), Modelling of Environmental Chemical Exposure and Risk, 245–249.

environment. The adequacy of models sufficiently depends on the validation of soil parameters, which describe the interactions of pollutants with soil components.

2. Empirical Analogies

There are some methods of receiving empirical data for modelling environmental processes. One of them is the method of empirical analogies. It is used for the investigation of different natural media and is not specific for soils. According to this the environment of some production plant is chosen as an object of investigation. Single and regular observations of the environment allow receiving the parameters of actual state of air, water, and soil. If the quantity (the dose) of pollutants, which have been discharged into environment, and the time of their entrance are known they may be used for the construction of a scheme of distribution of the investigated pollutants in time and in space in real and actual technogenic conditions. After that the results can be used for the analogous territories. The advantage of this approach is that the environmental situation takes place in reality, it is not theoretical and abstract.

There are some disadvantages of such approaches. They are the following: the information about the dose and time of entrance of pollutants is very often unknown for the researchers; it is difficult or may be sometimes impossible to fulfil the regular environmental observations during a long time; it is not easy to find fully analogous physico-geographical and soil-chemical situations. One of the difficulties consists of the fact that the list of factors which influence the environment is lengthy and it is almost impossible to take all of them into account and to calculate the dependence of environmental pollution from them.

3. Mass Balance

Another source of receiving parameters of soil pollution is the method of the balance calculation. Some types of information are used for this: a) technological, b) geophysical, c) the results of field modelling of soil pollution, d) the results of physico-chemical laboratory imitation of soil pollution.

Technological ones include the information about the type of production, the quantity and composition of waste, the regime of their entrance into the environment and so on. Geophysical approaches are based on the data about the direction of winds which dominate on the territory of investigation, about the distance of transportation of pollutants, size of solid waste particles, which include the pollutants and their composition. The limitation of these methods is the fact that soils are considered as an inert plane table, which only takes into account the pollution and nothing else. But this is not the case.

The peculiarity of soils consists of their ability to interact with pollutants, to transform them, to provide their migration down soil profiles and from soil to air, water, and plants.

The field modelling include the procedure of applying pollutants to the investigated soils. Some conditions are controlled in these experiments: the dose and composition of pollutants, the time of their application to the soil. The initial state of the investigated soils is recorded. The state of soils after some defined time and at a defined distance from the sources of pollution is controlled too. Many different soil characteristics may be determined in these experiments and they may be taken into account. The disadvantage of these models is the fact that they include soil as a «black box». The processes of transformation of pollutants in soils are not investigated; their results are only registered on the exit from the soil system.

The methods of physico-chemical modelling at laboratory scale are devoid of some of these disadvantages. During the laboratory experiment soil samples are pressed to the influence of the investigated chemicals. All conditions of interactions of chemicals with soils and soils properties are controlled, for example: the granulimetric composition of soils, moisture, temperature, reaction, the content of humus, soil minerals and some other components. But the laboratory experiment does not allow avoiding all difficulties in receiving soil parameters too. The problem consists of that soil is a heterogeneously polydispersed polyfunctional natural object. It means that many types of interaction of pollutants with different soil components take place in soils. The most important are the processes of adsorption-desorption, precipitation dissolution, ion exchange, the formation and dissociation of complexes. The list of soil parameters, which are informative for the environmental prognosis depends on the mechanism of transformation of pollutants in soil.

4. Limiting Processes

One of the ways to describe the physico-chemical situation in soil is the application of the principle of dominating or limiting process. This concept allows to substantiate theoretically the selection of one of the above mentioned processes of transformation of substances in soil as dominating, to imitate this process in the laboratory and to receive empirically those soil parameters which characterise the mechanisms of transformation of the substance in soil. This approach takes into consideration the characteristics of substances, the specification of the soil composition and the type of physico-chemical conditions in soils. Some examples of regularity were estimated when the laboratory modelling of chemical exposure was simulated on different soils in the laboratory experiments.

The principal difference between the types of interaction of organic and mineral pollutants with soils was estimated. It is connected to the biotic and abiotic mechanisms, which are involved in the transformation of organic pollutants. The participation of the biotic ones, however, is limited to soils, which are polluted with mineral substances. That is why the dependence of the environmental situation from temperature and moisture of soil, from the presence of available sources of nutrition for micro-organisms should be investigated when soils are polluted with organic substances. For example, the model, which was based on the parameters of decomposition of the pesticide «atrazin», was the most adequate among others. The climatic conditions, which provide the favourable level of heat and moisture for micro-organisms can influence this conclusion. In the northern regions the leaching of organic pollutants through the soil profile may play the role of the dominating process.

At the same time there are some examples where the biotic factors have provided a danger for the environment in the case of their pollution with mineral substances. It has taken place in soils and waters polluted with mercury compounds. One of the most toxic mercury species is methylmercury. The last is the product of microbiological activity and their quantity depends on the supply of easily available organic matter in soils and waters, which are necessary for micro-organisms.

The laboratory experiments have shown that the process of Zn-adsorption was the dominating one in the distribution of the element among soil solution and solid components in acid soils. The interactions of Zn-containing solutions with acid soils were investigated in static and dynamic laboratory conditions. The parameters of the equations of Langmuir and Freundlich were calculated and used for the estimation of Zn-adsorption to soils. Soil parameters such as the granulimetric composition of soil, the content of organic matter and clay minerals may be informative for predicting the metal migration in the profile of acid soils because they provide the quality and quantity of the active surface of soil particles were the processes of adsorption-desorption take place. But in soils with neutral and alkaline condition the process of precipitation-dissolution plays the dominating role in the distribution of metals in heterogeneous soil system. In this case the soil reaction but not the parameters of soil surface are informative for the prognosis of the migration of metals in landscapes.

5. Validation

The results of laboratory experiments were compared with the field ones. A field experiment was carried out that consisted of pouring Zn-containing solutions through the soil for one month. The lysimetric waters were collected and analysed. The results of two types of imitation of soil pollution, the characteristics of sorption ability of soil and Zn migration in the soil profile were compared.

The results of these two experiments have shown that the migration of metals in real forest ecosystems is more intensive than can be predicted on the base of modelling of the adsorption ability of soils which was determined in the laboratory experiments. It means that it is insufficient to use only soil adsorption characteristics for predicting the migration of metals in the landscapes. The type of soil, the structure and size of soil pores, cracks, holes of plants roots and soil animals, water transportation and other factors in the «living» soil of natural forest conditions influence the migration of metals and must be taken into account when the models of metal migration in soils are elaborated. Soils of the agricultural landscapes are more homogenised during the long period of their utilisation and the difference between the results of field and laboratory experiments is lower.

There are some experiments of modelling of environmental situations under acid rain exposure. The investigations were carried out in the field in conjunction with laboratory experiments. The subjects of investigation were acid soils of the south taiga of Russia. The experiments were based on the imitation of action of acid mixture on soil. It was estimated that the soil adsorption complex could provide the buffering capacity of the investigated soils to acids sufficiently, but not fully. It was supposed that the bases of the decomposed organic residues take part in the processes of neutralisation of acids rain too.

6. Conclusion

The adequacy of mathematical models of the environmental chemical exposure depends sufficiently on validation of soil parameters, which are used in these models. Soils receive the substance from technogenic sources and pass them to the environment as a secondary source of pollution. But the secondary flow of the substance has another composition than the initial one because of the interactions of pollutants with soil components, their transformation and migration in the soil profile. The validation of soil parameters for the modelling of the environmental situation may be provided only by the combination of theoretical investigations, laboratory and field imitation of soil pollution and analyses of the real state of the polluted soils.

RISK ASSESSMENT AND RISK MANAGEMENT OF INDUSTRIAL CHEMICALS IN POLAND

Country Report

JAN A. KRAJEWSKI
Nofer Institute of Occupational Medicine
Lodz
Poland

Abstract

The author describes the procedure used in Poland for exposure assessment and health risk management for workers exposed to chemicals present in work environment. It is shown how the recent changes in legislation have had impact on this procedure. Modelling is an important tool in predicting the concentrations of environmental chemicals and should be used on a larger scale.

1. Risk Assessment of Chemicals in Poland

Risk assessment and risk management of chemicals found in the work environment of Polish industrial plants is carried out using the standard US EPA procedure. The procedure comprises four stages: risk identification, determination of the dose-response relationship, risk assessment, and risk management.

Risk identification involves the collection and assessment of data on the exposed populations, exposure conditions and on the chemicals suspected of causing adverse health effects.

The stage of dose-response assay is omitted for chemicals, for which the permission to manufacture and to market it, has been already issued, since it had been already performed before the permission was granted.

In the work environment, workers are usually exposed to airborne chemicals in the form of gases, vapours and dusts. Generally, the exposure is through inhalation, although in some cases dermal and oral absorption is also possible. Exposure is assessed quantitatively by analysing air samples, collected from the workers' breathing zone, with the aid of e.g. gas or ion-exchange chromatography, atomic absorption spectrometry, or spectrophotometry. Quantitative assessment is also possible through the determination of chemicals or their metabolites in systemic fluids (urine, blood, saliva), or by the determination of the biological markers of absorption, or systemic response.

J.B.H.J. Linders (ed.), Modelling of Environmental Chemical Exposure and Risk, 251–253.

The risk can be assessed and characterised by combining the information and data obtained during the last three stages. The assessment is usually performed by comparing the determined exposure level with the binding occupational exposure limit.
Risk management comprises activities intended to reduce or eliminate adverse effects of chemicals on human health through the development of relevant regulations, their enforcement, as well as risk communication and information.
The Nofer Institute of Occupational Medicine is actively involved in performing all stages of the risk assessment and management process. Risk identification is supported by a number of databases containing information on: physical and chemical properties of chemicals; exposure under specific industrial conditions; toxicity to humans; hygienic standards in various countries. Use is also made of the database containing demographic data and information on population's health.
The dose-response relationship is determined for the chemicals when their toxicity has not yet been assessed. Toxicity is tested on laboratory animals, and the no-observed-adverse-effect-level (NOAEL) values reduced, by the value of the uncertainty or safety factor, are used to calculate the reference dose, i.e. the value of the 24-h exposure which does not produce a perceptible increase in the risk of adverse health effects in the population under study. This stage makes an extensive use of modelling which, when applied for suspected carcinogens, may be divided into two groups: statistical models which involve matching specified curves with the experimental data, and models based on the mechanisms of cancer development, such as e.g. one hit theory or multi hit theory. Models are also applied for the high-to-low dose extrapolation.
The risk management involves: a) development and enactment of legal regulations on admissible hygienic standards (a group of experts has been set up at our Institute, whose task is to determine maximum admissible concentrations of chemicals in the workplace); and b) enforcement of the legislation. The latter task is accomplished by the Sanitary and Epidemiological Stations. The risk management involves also training of various groups of people, from the exposed workers to the decision-makers. For this purpose, the School of Public Health has been founded at our Institute. Risk communication, which makes it possible to study risk perception and inform on activities which can be undertaken to reduce the risk is also essential in that respect.
The procedure outlined above relates to the existing chemicals. When an investor is going to start a new plant, or use a new process in an existing industrial plant, he is obliged to obtain permission from the local authorities. Environmental Impact Assessment (EIA) which rates the environmental effects of the project, constitutes the main prerequisite for granting the permission. EIA enables collection of data on the release of harmful chemicals to the occupational and, possibly, communal environments. Those data constitute particularly important criteria for granting (or refusing) the permission. When preparing an EIA document, use is made of modelling which involves, among other things, gas dispersion in the ambient air, or suspected concentrations of harmful chemicals in the water and soil.

2. Conclusion

The Polish regulations (the majority of which have been amended in this decade) oblige the investors to follow the risk assessment and risk management procedures only to some extent. Considering Poland's intention to become a member of the European Union, extensive work is conducted to harmonise our Polish regulations with EU legislation. The recent Act on Chemical Substances constitutes one of the major regulations on the marketing and production of chemicals in Poland. The executory provisions to this Act impose, among other things, the necessity to apply modelling for forecasting concentrations of chemicals in the air, water and soil.

MODELLING OF ENVIRONMENTAL CHEMICAL POLLUTION AND PERSPECTIVES OF EXPOSURE AND RISK ASSESSMENT IN UKRAINE

National report

LEONARD DOBROVOLSKY
Institute of Occupational Health
Kiev
Ukraine

Abstract

There is heavy pollution of environmental media in Ukraine by chemicals. Modelling of the distribution of pollutants in water and air is conducted by Ukrainian scientists. Instructions on calculation of chemicals in the water and air were worked out. In the nearest future appropriate modelling and epidemiological studies shall be conducted for assessment of chemical exposure and risk in Ukraine.

1. Introduction

In the present time annual anthropogenic impact on air quality of pollutant emissions from fixed sources is near 4.5 million ton and from mobile sources near 1.5 million ton in Ukraine. There are massive discharges of pollutants into Ukrainian water bodies. Annually the amount is more than 4.5 million ton of pollutants.
That is why working-out and implementation of mathematical models of environmental exposure to chemicals is very important as well from the scientific as from the practical point of view. The aim of this paper is to give a short review of the state and perspective of environmental pollution modelling in Ukraine as a first step of exposure modelling.
It is known that exposure of the population to pesticide is mostly by the oral route through food products. Ukrainian scientists have worked out a mathematical model called "pesticide-plant" [1,2].
In Ukraine, also an automatised control system of pesticides in agricultural and food products for 150 pesticides was worked out [3].
The elements of input information of pesticide into this load estimation model were:

- A real list of products and pesticides;
- The relative frequency of exposure (%) of each pesticide residue;
- The absolute values of pesticide residues in each product;
- The class of hazard of each revealed pesticide (in numbers);

J.B.H.J. Linders (ed.), Modelling of Environmental Chemical Exposure and Risk, 255–260.

- The maximum permissible level and ADI of the pesticide;
- The ration, proportion of food product.

2. Modelling Approach

Below the general formula of real and standard pesticide loads is represented:

$$\sum N_0 = \sum_{j=1}^{m} P_j \cdot \sum_{i=1}^{n} K_i \quad (1)$$

where:

$\sum N_0$	=	summary hazard of pesticide load in all products;
P_j	=	j-th product (j=1÷m) – number of products;
K_i	=	i-th coefficient of hazard (i=1÷n) – number of pesticides in concrete products

The evaluation of each product with pesticide residue was realised in an indirect way, through the hazard coefficient, which was calculated using the following formula:

$$K_i = \frac{C_1 \cdot R \cdot Q \cdot (R)}{C_2 \cdot 100} \quad (2)$$

Where:

C_1	=	the actual mean content of pesticide (mg / kg) in the product;
C_2	=	the maximal permissible level of pesticide in the product;
R	=	the frequency, with which a pesticide discovering in the product (%);
Q	=	the hazard class expressed in numbers;
R	=	the weight of product in the daily ration (kg).

If a standard coefficient (Ks) was calculated then substitute the ADI value for C_1:

$$ADI_{mg/man/day/food\ product} = \frac{ADI_{mg/kg/day} \cdot 60_{kg} \cdot 70}{100} \quad (3)$$

Using the already calculated value of $\sum N_0$, a calculation of the standard general value - $\sum N_{sg}$ was made. The ratio of both is characterised an integrated value of the load hazard index (J_i):

$$J_0 = \frac{\sum N_0}{\sum N_{sg}} \tag{4}$$

Where:

J_0 = the general integrated index, which is received on the basis of all products containing pesticide residues.

Under normal conditions the index $J_0 \leq 1.0$. A degree of hazard is determined when the index $J_0 > 1$ (3).

3. Model Implementation in the Ukraine

There were several schools in the former Soviet Union where environmental modelling and GIS were developed [4, 5, and 6]. In Ukraine more attention was dedicated to working out and using mathematical modelling of water pollution. Mathematical modelling can be distinguished into three classes: deterministic models based on differential equations, statistical models and imitative models.

3.1 DETERMINISTIC MODELS

Environmental media are very complex systems with many interdependent parameters. It is impossible to foresee many of them in deterministic models. They can be used only for solving particular local problems. It means that they must be built based on studies of physical, chemical and biological processes in the environment and reflect them in space and time. Its value consists in the obviousness of case-consequence relations in these processes.

3.2 STATISTICAL MODELS

Statistical models are models of the "black-box" type, which structure and parameters are determined on the basis of measuring information by minimisation of given criteria. Its quality consists in the simplicity and comparative less sensitivity to fluctuations of object studied.

3.3 IMITATIVE MODELS

The value of imitative models consists in the possibility of jointly using two previous model types as interdependent blocks in the general process of imitation. The feasibility of systematic development by improving some blocks make imitative modelling more perspective than the other two. In contrary to the first two classes of models, imitative model can in some way predict consequences of future loads and management strategy. The Ukraine-USA project "Water quality management for the Kaniv water reservoir" may serve as an example of modelling water pollution. The project included the development of a water quality management model with several on-site studies on model adaptation and verification and the development of a data bank for model programmers [7]. As far as in 1971 the well-known mathematician A.G. Ivakhnenko formulated tasks of modelling the ecological system of surface water [8]. Many mathematical models of water pollution were developed by scientists from Kharkov research institute on water protection [9, 10, and 11]. Intensive studies in this direction are conducted at the Dnipropetrovsk university [12, 13, 14, 15, 16], the Kiev institute of cybernetics and the technical university [17]. The Chernobyl accident have raised questions concerning the management systems of water pollution protection [18].

4. Outlook

The prognosis of changes of the ecological state of atmospheric dynamics is possible only on the basis of mathematical modelling. The results of prognosis constitute a basis for the development strategy of measures for improving the ecology situation. The method of mathematical modelling is only an instrument of the calculation and evaluation of probable ecological damage, a prognosis of air pollution resulting from an accident and other extraordinary events connected with discharges of pollutants into the atmosphere. The increasing demands to the accuracy of the prognosis raise from one side a problem in the developing process.

Rational mathematical models, taking into account the influence of various process in air media, and from another side – making effective, reliable, flexible computer technologies, realised on personal computers and adapted for a broad use in practice.

Ukrainian scientists proposed three levels of modelling:

The spatial transfer in the atmosphere on the scale of distribution of several kilometres. This is a first order mathematical model, and on its basis the program complex "Prognoz-1" was worked out.

For the description of distant air pollutants transfer (several hundreds of kilometres), the equation used averaged possible pollutant distribution heights in the atmosphere. This is a second order model. On this model basis the program complex "Prognoz-2" was developed. Modelling of atmospheric pollutants of distant transfer processes was realised on the example of the distribution of cities and industrial centres of discharge of one of largest Ukrainian industrial regions: Dnipro-Donbass [17].

There are instructions concerning the calculation of the concentration of harmful substances and their distribution in the atmosphere [19, 20, 21] and some examples of modelling of the air pollution in the industrial areas [,22].

5. Conclusion

So inspite the perceived risk from chemical pollution in Ukraine is high, the actual and relative risks are unknown due to lack of capacity to measure low levels of toxins in

environmental media and inexperience with using environmental data and behaviour information to assess exposure and determine relative risks.
Development of appropriate models, scenarios and epidemiological studies in the nearest future can help to assess the exposure of various environmental media and determine risks.

6. References

1. Ivanova L. N. (1971) A synthesis of mathematical model of process transformation and transfer of pesticides in the soil-plant system. In L. I. Medved (Ed.), *An application mathematical methods for estimation and prognosing of pesticides real hazard.* VNIIGINTOX, Kiev, pp 27-28.

2. Spynu E. I., Ivanova L. N. (1977) Mathematical prognostication and prophylaxis of environment pesticides pollution. Meditsina, Moscow.

3. Kuchak Yu. A., Tkachenko T. N., Dutka M. A. (1988) Modelling of real pesticide load hazard on the basis of territorial automatized control system data bank. In A. V. Pavlov (Ed.), *Hygiene of use, toxicology of pesticides and polymer materials.* VNIIGINTOX, Kiev, 18, pp 36-39.

4. Modelling and prognostication in ecology (1978). A. M. Mauryn (Ed.). Latvian univ., Riga.

5. Modelling of substance and energy transfer in the natural systems (1984). V. K. Arguchintsev (Ed.). Nauka, Novosibirsk.

6. Modelling of the environment (1986). A. M. Trofimov (Ed.). Geographical soc. USSR, Leningrad.

7. Kachiashvili K. I. (1990) Modelling of analytical instruments and processes of water media pollution. Society "Knowledge", Kiev, 1990.

8. Ivakhnenko A. G. (1971) A formulation the task of water ecological system modelling,*Avtomatica*, 4, pp 20-23.

9. Yeryemenko E. V., Kolpak V. Z. (1973) Calculation of concentration of passive admixtures in the river with tributaries. In Lozansky V. R. (Ed.) *Problems of waters protection* VNIIVO Kharkov, 4, pp 144-153.

10. Byelogrirov V. P., Vasilenko S. L. (1980) Statistical models of quality water changes. In Lozansky V. R. (Ed.) *Quality of natural waters management.* VNIIVO Kharkov, pp 61-69.

11. Yeryemenko E. V. (1980) Mathematical modelling of water quality forming with the aim of management and planning water protection. In Lozansky V. R. (Ed.) *Quality of natural waters management.* VNIIVO Kharkov, pp 22-30.

12. Belyaev N.N., Khrutsch V. K., Kuvshinov V. I. (1995) Mathematical simulation of the water environmental pollution// Geological studies: state and perspectives. Collection of sci. works of int. sci. conf. Kyiv, p. 52.

13. Khrutsch V. K., Belyaev N. N. (1994) Mathematics modelling of pollution transfer in Dnipro water basin // Numerical methods in hydraulics and hydrodynamics: Rep. abstracts on the 1st int. conf. Don. Univ., Donetsk, p. 17.

14. Belyaev N. N. Numerical simulation of the pollutant transfer in the rivers. In Prishyakov V. F. (Ed.) *Sustainable development, environmental pollution and ecological safety.*- Dnepropetrovsk.-State Univ. – 1995, 1, p. 166.

15. Belyaev N. N. The mathematical simulation of the pollution of the ground waters // Intern. Congress "Water: Ecology and Technology" Moscow, Sept. 6-9. –1994, 1, p. 90.

16. Belyaev N. N. et al. The prediction of the River Dniepr pollution by the method of the mathematical modelling // Ibid., p. 91.

17. Zgurovsky M. Z.,Skopetsky V. V., Khrutch V. K., Belyaev N. N. (1997) Numerical modelling of environment pollution. Naukova Dumka, Kiev, p. 377.

18. Plesunov V. M., Yakovlev E. A. (1991) Conception "Dniepr": determination and choice of effective management system of radioactive and toxic substances migration processes in the limits of water collecting River basin. In. *Radiological, economical and legal aspects of land-tenure after accident of the Chernobyl atomic power plant.* Materials of the 1st sci. conf., Kiev, March 27-29, 1991, pp 31 - 33.

19. Berlyand M. E., Koshkin A. N., Onicul R. I., Eds. (1975) Instruction on calculation of dispersion in the atmosphere of harmful substances contained in the enterprise discharges. SN 369-74, Strogizdat, Moscow.

20. Method of calculation of harmful substances concentration of enterprises emissions in the atmospheric air. (1987). Gidrometeoizdat, Leningrad

21. Tischenko A. A. (1991) Protection of atmospheric air. Calculation of harmful substances content and their distribution in the air. Reference book. Khimiya, Moscow.

22. Belyaev N. N., Khrutsch V. K., Kuvshinov V. I. (1995) Modelling of the air pollution in the industrial areas // Geological studies: state and perspectives. Collection of works of Int. scientifical and practical conf. – Kyiv – p. 75.

OVERVIEW ON ENVIRONMENTAL SITUATION IN ALBANIA AND SOME ISSUES IN THE FIELD OF "MODELLING"

Albanian experience

TANJA FLOQI, QECAMEDIN KODRA, GENC LUARASI AND BUJAR REME
Institute of Studies and Design for Light Industry
Rruga "M. Gjollesha" 56,
Tirana,
Albania.

Abstract

Geographical data and socio-economic data during the transition period of Albania are presented together with the current institutional and legislative framework. Projects carried out in Albania may be financed by the government on one hand and by international bodies, like PHARE, on the other hand. More and more data are becoming available concerning industrial branches and their discharges into the environment. The water supply, the quality of drinking water and its health impacts receive a lot of attention currently in Albania. Finally, some issues in the field of modelling experience and an outlook into the future are dealt with.

1. Introduction

1.1 GEOGRAPHY

Albania is a small country situated on the eastern shore of the Adriatic Sea, with former (ex-)Yugoslavia to the north and east and Greece to the south. The total land area covers 28748 km^2, of which 76% are mountains or hills.
Albania's climate is Mediterranean with variable precipitation and temperature throughout the year, with hot dry summers, and generally mild winters with abundant rainfall.

J.B.H.J. Linders (ed.), Modelling of Environmental Chemical Exposure and Risk, 261–270.

1.2 POPULATION

Albania has a population of approximately 3000000 inhabitants. The population has doubled over 37 years. It is considered a young populations. Its average age is 28.6 years.

1.3 NATURAL RESOURCES

Albania has many natural resources: copper, iron-nickel, chromium ore; petroleum and natural gas, coal, etc. It is one of the world's largest producers of chromium. It has hydropower stations to support 80% of its total electricity productions.
In today's international categorisation of countries, Albania is included amongst those economies that are described as being "in transition" from a centrally planned to a free market one.

2. Specific Items on Albanian Transition

2.1 MIGRATION

Beginning from the year 1990 the Albanian population got familiar with new phenomena, which impact its structure and growth to quite an extent.
It is foreseen that by the year 2010-2015 an intensive process of internal migration will determine a concentration of population in the coastal region. Thus, the area Tirana - Duress (respectively, capital and the greatest port of Albania) will have about 1.5 million inhabitants, that means nearly half of the entire population.
Emigration was virtually non-existent in Albania until 1989. It is estimated that the number of emigrants still continues to increase arriving about 400.000 in Greece, nearly 100.000 in Italy, etc. A good part of the emigrants are young and highly educated thus affecting the reproductive capacity and the intellectual potential of our country.

2.2 BOOM OF TRAFFIC AND CONSTRUCTION ACTIVITIES

The data in Table 1* below show clearly the extraordinary growth of the mobile means of transportation (especially private) during the period 1990-1999.

Table 1. Growth of public cars

Year	Cars
1990:	20.000
1992:	40.000
1993:	60.000
1994:	80.000
1996:	100.000
1999:	180.000

* *These figures are and approximate values and not official.*

A lot of factors such as: the rapid increasing of the number, the obsolete state of engines, the use of bad quality oil, damaged roads, etc. gave reason to consider the situation as "explosive".

3. Institutional framework

3.1 THE MAIN BODIES

National Water Council (NWC) is a committee of the national government. The NWC was set-up by Presidential decree in 1994. Only one person has been appointed to its Technical Secretariat (TS). The director of TS is Franco SARA based in the Ministry of Public and Transport.
NWC has the overall authority to decode water protection and management strategy; NWC has the responsibility for establishing regulatory structures, standards and requirements for water management and protection.
National Agency of Environmental (NAE) has the overall authority to protect the environment: (i) implements the Environmental Impact Assessment (EIA) for activities having a strong impact on the environment and which are particularly dangerous to human health (ii) issues surface water), (iii) supervises environmental monitoring by various governmental institutes and collects, files and processes the data, (iv) Regional Environmental Agencies fall under the jurisdiction of NAE.
Ministry of Health (MH). The Directorate of Primary Health Care and the State Sanitary Inspectorate is responsible for the control of the quality of drinking water with the laboratories of public health directorates in the districts, the Ministry of Health regulates and supervises scientific institutions responsible for water quality monitoring. MH proposes the regulations, standards and procedures for the control of water quality in consultation with National Water Council (Technical Secretariat).
Institute of Public Health (IPH), replacing Institute of Hygiene and Epidemiology, it is in charge of the monitoring the level of air pollution, of the water quality in urban and rural areas and industrial centres. IPH is also in charge of monitoring and control of soil pollution in the whole country.
Ministry of Public and Transport, General Directorate of Water Supply and Sewerage (GDWSS) is responsible for water supply, wastewater treatment and disposal. GDWSS is also responsible for the investment in this domain and for the co-ordination of investment by the donors. It prepares, proposes regulations, standards and supervise the projects for the water supply network and the wastewater treatment plan in the country.
Ministry of Public Economy and Privatisation, Sector of Safety and Environmental Protection, several institutions are part of this ministry:

- *Albanian Elctroenergetic Corporation*
- *Institute of Metallurgical Research &Design*
- *Albanian Geological Survey*
- *Alb - Chromium*
- *Alb Crupper*
- *Albpetrol*
- *Institute of Extracting & Processing Minerals Technology*

- *Research & Design Institute for Industry*
- *Chemical Techonology Research Institute*

Ministry of Agriculture and Food, Directorate of Irrigation and Fisheries has now the authority to manage and control irrigation schemes and channels, but will be transferred to the NWC.

Institute of Soil Study is monitoring the level of soil pollution and the level of contamination of irrigation water.

Institute of Hydrometeorology (IHM) was founded in 1962 under the Academy of Sciences, replacing the Hydrometeorology Service that was created in 1949. The IHM is a research institute, responsible for the collection and analysis of all meteorological and hydrological data. Its task includes the management of all pluviometric and climatological stations, the observation of water levels, the measurement of river discharges and the sampling and analysis of water in the rivers and other water bodies to assess its quality. IHM is in charge of monitoring and control of the quality of surface water, groundwater and rainfall in Albania.

Institute of Nuclear Physics is in charge of monitoring the radioactivity of the environment and transboundary waters.

4. Legislative Framework

Several laws came into force in Albania already. An overview is presented below:

- Decree No. 5105, dt. 30/10/1973 "On environmental protection"
- Decree No. 7452, dt. 05/01/1991 "National Committee on Environmental Preservation and Protection"
- Law No. 7643, dated on 2/12/1992 "On state Sanitary Inspectorate".
- 1993 The law No. 7664, dt. 21/01/1993 "On environmental protection"
- Decision No.26, dt. 31/01/1994 "Hazardous wastes and residues"
- Decision No. 541. dt 25/09/1995 "On Ministries, institutions and legal persons charged for monitoring and environmental control"
- Law No. 8102. Dt. 23/03/1996 "For regulator framework of water supply and removal sector".
- Law No. 8094, dt. 21/03/1996 "On public waste disposal"
- Law No. 8101, dt. 28/03/1996 "On taxation system of the public waste disposal" Amended with Law Nr. 8338 date 30/04/1998.

In addition several projects were funded by government and by international organisations.

1. Projects financed by government.
    - Inventory of hazardous chemicals and their disposal.
    - Assessment of pollution caused by chemical processing industry and mitigating measure.
    - Soil pollution in suchetibillity aeries.
    - Survey on environmental impact of oil discharges.
    - Monitoring for water quality and level pollution of surface waters.

2. Phare projects
   - Institutional strengthening of E.P.C.
   - Equipment for REA-s (local level).
   - Regulatory framework.
   - Public Awareness.
   - Management plan on National Park of DAJTI
   - National Water Strategy.
   - Project & design for sewage water treatment plant in Vlora.
   - Project & design for sewerage water treatment plant in Pogradec.
   - Management plan on Karavasta lagoon.
   - Urban solid waste Management.

5. Monitoring programs

The monitoring began in 1996 and functioned regularly till 1991. In the network of air quality control there were inoculated 23 stations in 10 cities. All stations were in urban areas, expect one in RIHE (Research Institute of Hygiene and Epidemiology).
After an interruption of a few years period, the programme of air pollution monitoring restarted using passive methods. A new automatic station is installed in the RIHE (presently, IPH - Institute of Public Health). The monitored pollutants are: O_3, NO_x, CO, SO_2, PM10, etc. Different surveys, free inspections and people's complaints confirm the presence and effects of pollutants.

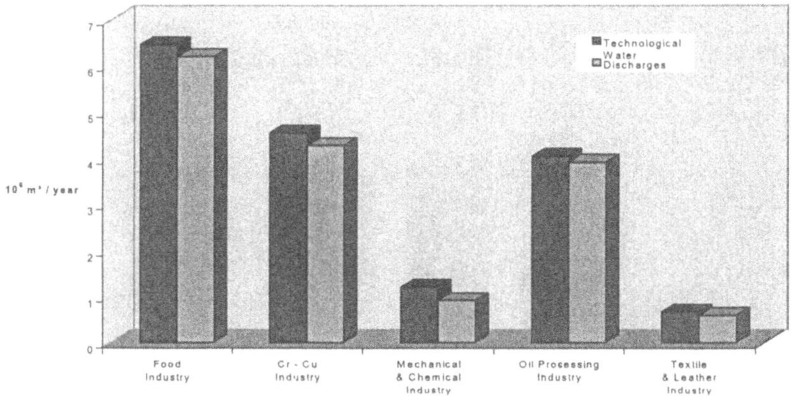

Figure 1. Technological Water and Discharges in Albania

In Figure 1 the amounts of technological water and the total discharges in Albania are shown for different types of industry. The units are in millions of cubic metres per year.

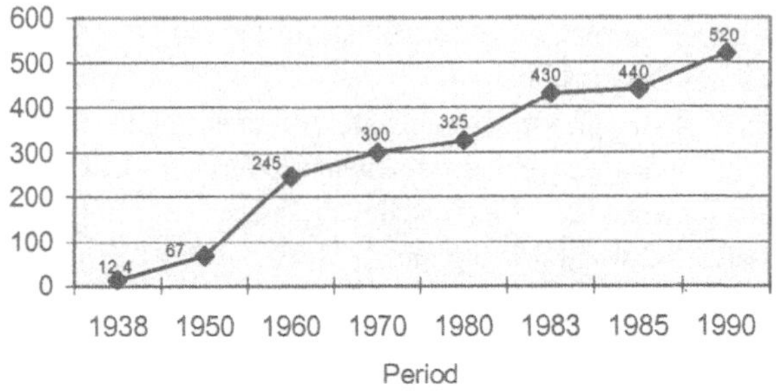

Figure 2. *Water amount (litre/capita/day) according to the production from pipelines of the cities*

Figure 2 shows the increase over the years of the amount of water used by the inhabitants according to the production capacity of the waterworks of the cities, while figure 3 gives an overview of the increased amount of pipelines for public water supply constructed over the years.

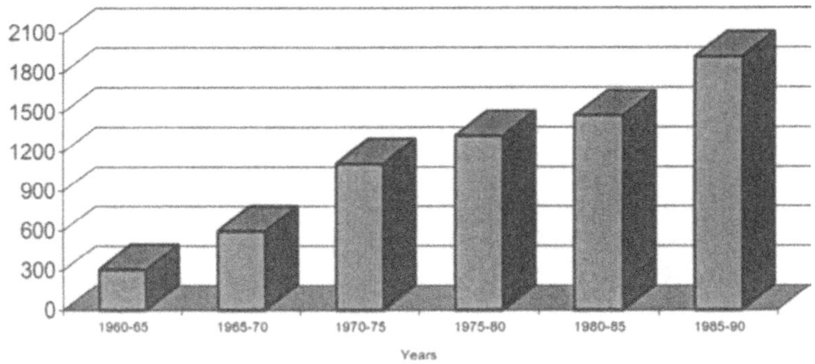

Figure 3. *The construction of water pipelines in the country side 1960-1990*

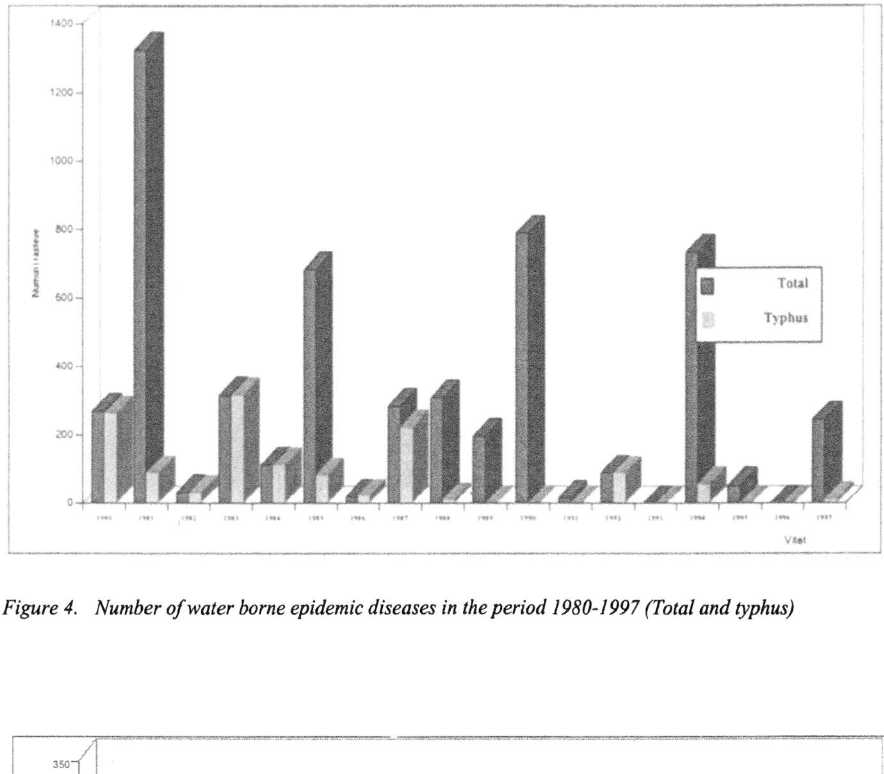

Figure 4. Number of water borne epidemic diseases in the period 1980-1997 (Total and typhus)

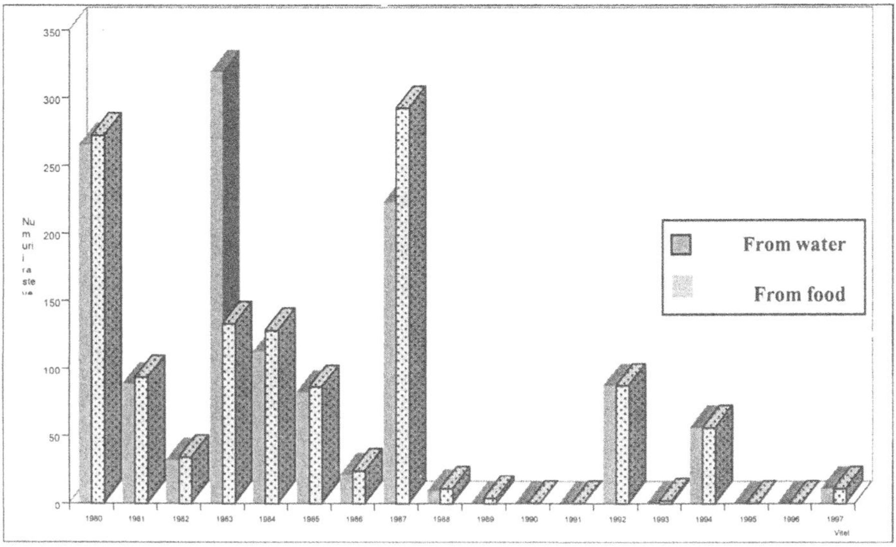

Figure 5. Number of abdominal typhus from water and food origin in the period 1980-1997)

TABLE 2. The level of water pollution along the coastal line

Samples sites	Unit	Cr	Fe	Ni	Cu	Zn	Pb	U
Velipoje	μg/l	0.4	1.8	1.8	1.6	3.5	0.3	4.8
Shengjin	μg/l	4.8	3	1.9	2.7	3.1	1	3.5
Patok	μg/l	0.5	3	2.2	3.1	3	0.8	3.8
Porto Romano	μg/l	1.4	12.8	2.3	16.5	8.5	2.3	3.9
Durres	μg/l	1.5	58.3	1.6	4.8	50.5	1.8	3.3
Iliria	μg/l	1	4.8	2	6.2	5.8	2.7	4.3
Golem	μg/l	1.6	5.5	1.4	11.5	9.2	3.8	3.2
Vlore	μg/l	0.8	3.5	1.3	1.9	5.5	1	3.4
Himare	μg/l	0.7	2	0.2	1	9.2	0.5	4.5
Borsh	μg/l	0.7	3.5	0.3	2	1.7	1.2	4
Sarande	μg/l	0.6	4.5	0.4	3	3.5	1.5	3.9

TABLE 3. Some qualitative indicators of Albanian lake waters (mg/l) for year 1996

Lake	Sta-tion	Depth (m)	TOC	'pH	Ca	Mg	HCO_3	Cl	SO_4	Suspended solids
Prespa	1	0	23.0	8.65	10.02	12.6	122.0	10.63	10.0	2.2
		15	23.0							
Ohri	1	0	23.0	8.67	16.03	3.40	122.0	10.63	8.0	0.31
		50	7.8							
		75	6.8							
		94	6.8	8.29	16.03	4.86	122.0	10.63		
Shko dres	1	0	26.0	9.49	14.28	3.40	122.0	10.63	15.0	3.10
		6.5	25.9							
	2	0	26.5	8.99	14.28	3.40	122.0	10.63	14.0	4.20
		5.0	26.0							
	3	0	26.0	9.12	10.02	4.25	91.5	14.18	12.0	3.00

(Hydrometeorological Institute)

TABLE 3. (Continued) Some qualitative indicators of Albanian lake waters (mg/l) for year 1996

Lake	Sta-tion	Depth (m)	Dissol-ved O_2	NKO	NBO	NH_4	NO_2	NO_3	SiO_2	Total P
Prespa	1	0	7.21	0.32	0.94	-	-	-	10.0	2.2
		15	23.0							
Ohri	1	0	9.80		2.08	-	-	-	8.0	0.31
		50	9.99		2.55					
		75	8.69		1.05					
		94	9.25		1.00	-	-	10.63		
Shko dres	1	0	7.58	0.80	0.92	-	-	10.63	15.0	3.10
		6.5	7.13		0.66					
	2	0	7.58	0.92	0.76	-	-	10.63	14.0	4.20
		5.0								
	3	0	12.54	0.92	4.22	-	-	14.18	12.0	3.00

(Hydrometeorological Institute)

The figure 4 shows the total amount of water borne diseases in Albania in the period 1980 – 1997 compared to the amount of typhus incidents, while figure 5 shows the same typhus related incidents compared to the amount originating from food in the same period 1980 – 1997.
Table 2 gives an overview of the pollution situation along the Albanian coast with respect to some heavy metals. In addition, some water quality indicators of Albanian lake waters are presented in table 3.

6. Modelling Activities

In order to give an overview of the real "modelling" in Albania we have carried out a number of consultative meetings with specialists of various institutions, interested in modelling as:

- *Institute of Public Health,*
- *Albanian Geological Survey*
- *Albanian Elctroenergetic Corporation*
- *Institute of Metallurgical Research &Design*
- *Alb - Chromium*
- *Alb Copper*
- *Albpetrol*
- *Institute of Extracting & Processing Minerals Technology*
- *Institute of Studies and Designs for Light Industry*
- *Chemical Technology Research Institute*
- *Institute of Soil Study.*
- *Food Research Institute*
- *University of Tirana*
- *Institute of Hydrometeorology*
- *Institute of Nuclear Physic*

7. Conclusion

As the conclusion we can say that:
The "modelling" is scarcely known and consequently is scarcely used (hydrodynamic modelling, etc.). The term "modelling" is not encountered in development programs or in the preliminary projects/studies. Almost all institutions carry on the mathematical processing of the data taken from the observations and studies.
The participation in this workshop will be a good opportunity to obtain information regarding the values of the methods on the application fields of "modelling".
A program is necessary in order to activate the modelling in Albania.
Please, take in account our difficulties in providing and in covering the necessary dates, which are to be used in the modelling (chemistry, demography, hydrogeology, meteorology, etc.).

Nevertheless we can not avoid an initial classical scheme of programming.

a. Studying phase:
   - Theoretical know how (from literature)
   - Establishment of team/teams, with the specialists of various fields

b. Training phase:
   - Terming through intensive courses combined with practical illustration
   - Needs identifications & formulation
   - The modelling establishment in the context of EIA

c. Implementation phase.

CONCLUSIONS AND RECOMMENDATIONS

I. Conclusions

Advantages and disadvantages of modelling and risk assessment

Advantages

- Hazard/risk assessment systems are based on scientific principles.
- Hazard and risk assessment can be powerful tools in assessing consistently the behaviour, effect, hazard and risk of chemical substances with regard to man and the environment.
- Common language and terminology is provided between scientists, decision makers and the public.
- Risk assessment is a controllable process (GMP).
- Risk assessment is a possibility for comparison with other natural risks.
- Risk assessment is a decision support tool and can be used for registration of chemicals in the environment through estimation of limit values.

Disadvantages

Technical objections:

- inaccuracy of our knowledge concerning:
  - properties of pollutants;
  - magnitude and type of emission;
  - environmentally variable parameters;
- uncertainty in exposure assessment;
- different safety factors in ADI values.

Non-technical objections:

- numbers (risk values) are emphasized out of all proportion;
- risk values are simple for communication but complicated for understanding;
- overestimation of expertise might push in the background social, cultural factors and subjective observations.

Concluding remarks

- Continuous improvement and refinement and validation is essential;
- Hazard and risk assessment are supporting decision making but cannot replace independent scientific judgment;
- Taking the above into account modeling can be powerful tools in assessing risk.

Comparative assessment of models and their use in representative regions

- Comparisons should only be made if the models have the same purpose;
- Methodology:
  - compare models with measured data (model as a black-box);
  - compare the scientific background.
- Decision making:
  - a comparative assessment should be based on a considerable number of results;

  - availability of input parameters is important (simple model with available data vs complex model without data);
  - what to do with model B and C if the comparative assessment shows that model A is the most favourable?
    - recommend not to use B and C? Very difficult due to preferences of individuals and countries;
    - develop new model which combines the strengths of A, B and C? funding is a potential problem in developing new models.

II. Recommendations

Recommendations for the use of models for specific environmental compartments

- Priorities for use should be based on the areas of use.
- There are two major types of use for modelling – regulation and research.
- It is not possible to recommend one specific model but some areas can be highlighted.
- Regulatory models should be able to accurately predict concentrations in specific environmental compartments.
- Pesticide models are generally more advanced than models for industrial chemicals.
- The selection of appropriate model inputs is determined by the modelling objectives. Mean values provide "best estimates" while worst-case inputs provide "worst-case" estimates.
- Probabilistic modelling is useful but requires knowledge of the distributions of input paramaters. The statistics of extreme values is a specialized area of research.
- Fuzzy models could be used to estimate the range of uncertainty.
- Generally, the first priority in modelling is to represent the key processes. The second priority is to obtain distributions of input parameters.
- More development is needed for ecosystems, air (both short range and long range transport) and the terrestrial compartment.
- There needs to be international coordination of further development of key models. More cooperation is needed to develop internationally recognized models.
- The role of FOCUS is to recommend specific models for use in pesticide registration. These groups should continue to be supported.
- User support should be provided for models. More training is needed to build the expertise of regulators.
- Models should be updated based on current scientific developments.

Recommendations for future collaboration in the validation of models

- Validation is fitting model results with reality. One of the goals is to validate the inner structure of the model (i.e. the equations). Internal versus external validation:
  - input parameters should only be used with validated regression ranges

  - the user should ensure that the model is appropriate for the chemical being modelled
  - the theory used in modeling should be based on laboratory and/or field experiments
- The decision to accept or reject validation work should be based on the purpose of the modeling and the scientific consequence.
- Spatial and temporal resolution can be a problem. Quality criteria need to be developed for measured data
- Measured concentrations should be clearly defined. Are they representative of localized values or regional values? Are they mean or median values? What was the detection limit?
- Recommendations should be developed for determining the uncertainty in modeling results.
- Models require good measured values.
- Laboratories should have good quality control.
- It may be useful to develop a quality score for measurements.
- Databases are needed for environmental data, toxicity data and emission data.
- Typically, more data than just the concentration are needed for modeling. Additional parameters include, e.g. pH, temperature, location.
- There should be more collaboration on the validation status of certain models. Who should decide if a model is validated? Should it be expert judgment or up to an established group of modelers such as FOCUS? Perhaps NATO should establish an expert board for modelling.

Recommendations on future research needs on modelling

- Announce a small workshop to plan further research on modelling.
- Inform officials in EU 5th framework program of the decisions from this workshop and to help in the future development of future research programs.
- Most leaching models simulate either the top 1-2 meters of the soil profile or the entire unsaturated zone. The capabilities of these models should be extended into the saturated zone.
- Additional models are needed for specific problems such as rice, algal bloom control and forestry.
- Actions are needed to promote use of modelling in various countries, training of users, correct use by decision makers.
- Model(s) are needed for indoor air quality and formation and decline of metabolites in the environment. Also pharmacokinetic modeling is needed to translate external exposures into internal doses to more accurately assess risk.
- Pesticide exposure models need to be adapted and validated to address EU conditions. Additional site-specific scenarios should be defined and characterized to represent a wide range of geographic settings.
- Should introduce concept of a model passport which would make models more understandable. More training should be provided to ensure correct understanding

and application of models. Model developers should distinguish between dynamic and state variables and clearly state how parameters are validated and updated. The exchange formats for I/O files should be standardized to facilitate coupling of models and future refinement.

- Use more experience from ecological modelling.
- Uncertainty of model outputs should be expressed by using probabilistic approaches. Output values should be expressed scientifically (with ranges). Safety factors should not be incorporated into concentration predictions but kept separate to clearly indicate the scientific and regulatory contributions to the predicted values.
- The capabilities of time-dependent sorption and depth dependence of degradation should be added to leaching models.
- Not enough is known about how sorption and degradation influences movement via preferential flow.
- Need to improve the accuracy of estimating potential and actual evapotranspiration since these values dramatically affect recharge.
- Models deal reasonably well with chemicals, which are deliberately spread in the environment. We need to improve our knowledge of the environmental distribution of industrial chemicals.
- Scenarios should be developed for a wide range of geographic settings. Some countries have well developed modelling capabilities and scenarios while others have low capability but also significant pollution problems.
- Need to organize some means of information exchange on modelling.
- There is a research need to improve modelling capabilities of simulating simultaneous exposure from soil, water and atmosphere. Risk assessments should consider multiple contaminants and possible synergism.
- Fundamental research must be increased with a focus on more appropriate environmental measurements and how to express uncertainty.
- There is a lack of information in the EU on appropriate tiered assessment schemes for toxicology and ecotoxicology.
- More research is needed on methods for model validation. Formal methods need to be developed to assess model accuracy.
- Need for develop standardized scenarios for southern and northern Europe.
- Validation studies should be performed at a local level.
- More training needs to be provided for model users to ensure correct understanding and use of models.
- Model users would like to find a standardised package of validated and common models for use in exposure assessment and risk assessment.
- Support should be provided to countries that do not currently have internal modelling capabilities including training, making models flexible enough to accept regionally appropriate inputs and providing assistance on the consequences of regulatory decisions based on modelling results.

- To stimulate additional thought concerning interpretation of modeling results, training should be organized at an advanced institute in which young researchers are invited to collaborate and begin using modeling to address specific needs.
- Some parameters are used in models which are not part of pesticide registration packages. More research is needed to clarify the needs for additional data.
- Need to develop more geographically-referenced (site specific) modeling.
- Need for better understanding and tools to assess chronic air exposure for human health.

General recommendations

- Need to develop ways to express uncertainty in the results of models. Need to provide a more systematic way to understand consequences of uncertainty in input parameters.
- Need to develop improved approaches to validate models.
- Need to address how to handle simultaneous exposures from multiple chemicals and simultaneous routes of exposure (probably less important). The current assumption is that multiple simultaneous contamination pathways are possible but exposure is generally dominated via one chemical and via one pathway.

Specific recommendations

- Need more site-specific scenarios. Some data layers have been developed for all of EU and may be usable for pesticide modelling. In air acidification modeling, RIVM provides a common EU model for all countries which is adapted to various national and regional assessments. Countries should defend locally appropriate parameters rather than specific models. Would also help to address uncertainty or variability if a reasonable range of representative scenarios were available in each country.
- Probabilistic modelling developed and used to reduce the use of safety factors.
- To obtain the benefits available through modelling, steps should be taken to promote training, develop standardized scenarios, use of websites and exchange of software.
- More specialized workshops are needed to provide information and training in the use of modelling. Web sites could be provided to provide technical assistance and to distribute models.

What can we do to push these conclusions?

- A short course on environmental risk assessment and modelling will be offered in early 2000 in the UK focussed primarily on pesticide fate modelling using MACRO and PELMO.
- Mailing lists are available for some types of models.
- Country-specific scenarios are important for human exposure to pesticides because practices vary with each country. External exposure varies with country. However, internal distribution is constant with humans. Risk tolerance can vary regionally. Co-pollutants also vary regionally and affect human response.